Meta-analysis in Clinical Research: Principles and Procedures

Anoop Kumar

Meta-analysis in Clinical Research: Principles and Procedures

Springer

Anoop Kumar
Department of Pharmacology
Delhi Pharmaceutical Sciences and Research University (DPSRU)
New Delhi, India

ISBN 978-981-99-2369-4 ISBN 978-981-99-2370-0 (eBook)
https://doi.org/10.1007/978-981-99-2370-0

This Springer imprint is published by the registered company Springer Nature Singapore Pte Ltd.
The registered company address is: 152 Beach Road, #21-01/04 Gateway East, Singapore 189721, Singapore

Dedicated to my parents

प्रो. रमेश के. गोयल, पीएच.डी.
कुलपति
Prof. Ramesh K. Goyal, Ph.D.
FIC, FIACS, FAMS, FNASc, FIPS, FICN, FSCH, FISVPT
Vice Chancellor

दिल्ली भेषज विज्ञान एवं अनुसन्धान विश्वविद्यालय
राष्ट्रीय राजधानी क्षेत्र, दिल्ली सरकार
भारत का पहला फार्मेसी विश्वविद्यालय
(2008 के 07 अधिनियम द्वारा स्थापित)

DELHI PHARMACEUTICAL SCIENCES AND RESEARCH UNIVERSITY
Govt. of NCT of Delhi
The First Pharmacy University of India
(Established Under Act 07 of 2008)

Foreword

Advancement in medical science is mainly reflected through the search data compilation and publications. There is a plethora of literature and clinical data available for diseases that many times confuses clinical practitioners and scientists for the clinical decision-making process.

Meta-analysis is one of the innovative methods of quantitative analysis that helps in clinical decision-making, particularly in complex diseases. The results of the individual studies can be combined and integrated using suitable statistical procedures. A well-conducted meta-analysis can provide a more precise overall estimate along with a confidence interval. It also explains heterogeneity among individual studies, the effects of outliers on the overall estimates measures, and publication bias. This process although appears simple but is complex and requires the development of skills for data analysis and biostatistics coupled with advanced clinical knowledge. Although, there are several systematic literature review articles are available, a comprehensive book incorporating different facets of meta-analysis is not available.

The book entitled "Meta-analysis: Principles and Processes" by Dr. Anoop Kumar has not only given a simple yet complete view of meta-analysis with the methodological approach required in the field of medicine. I am confident that this book will serve as a guide for healthcare professionals, researchers, academicians, and regulators. The author and the publishers need to be commended and congratulated for providing useful information.

5/5/23
Prof. Ramesh K. Goyal

Prof. Ramesh K. Goyal
Vice Chancellor
Delhi Pharmaceutical Sciences And Research University (DPSRU)
(Govt. of NCT of Delhi)
Pushp Vihar, Sector-III, M.B. Road
110017

Preface

Meta-analysis is one of the well-known approaches, particularly in the medical field which analyse the data from individual studies using statistical methods to help physicians, regulators, and consumers to make better clinical decisions particularly, on controversial topics. The popularity of meta-analysis is increasing year by year and the availability of user-friendly software helps in increasing the number significantly. However, this analysis should be done carefully.

This book intends to bring more rigorous knowledge on meta-analysis. This book provides a general introduction to meta-analysis, steps involved in the process of systematic literature review (SLR), quality assessment, extraction, and analysis of data. Further, this book also explains the meta-regression, selection of model, determination of publication bias, and heterogeneity. The common mistakes, issues, and challenges are also described.

This book has been orgnised within 14 chapters, presented in a logical sequence. Chapter 1 is Introduction. Chapter 2 discusses the Systematic Literature Review (SLR). Chapter 3 discusses the Quality Assessment of Studies. Chapter 4 compiles the information regarding the Extraction and Analysis of data. Chapter 5 discusses Models in Meta-analysis. Chapter 6 discusses the Heterogeneity and Publication Bias in Meta-analysis. Chapter 7 discusses bias in meta-analysis. Chapter 8 provides an overview of Sensitivity and Subgroup Analysis. Chapter 9 discusses the Meta-regression. Chapter 10 discusses the Plots in Meta-analysis. Chapter 11 provides an overview of the Network Meta-analysis. Chapter 12 discusses the Registration and Software. Chapter 13 briefs about the Common Mistakes. Chapter 14 provides an overview of Challenges. Chapter 15 discusses Future Perspectives.

The author wishes to express their considerable appreciation to the publisher who took over the management of the production of this book in difficult circumstances and whose contribution is much appreciated. I am always thankful to my parents, my younger brother (Arun Kumar), my wife (Dr Ruchika Sharma) for their constant support throughout my life. Special thanks to my son (Master Nivaan Sharma) who has always relieved all my tiredness with his innocent things.

I trust this book will be useful to researchers, faculty members, and medical and pharmacy students who would like to start a SLR and meta-analysis.

New Delhi, India Anoop Kumar

Introduction

The present book set of Meta-analysis: Principles and Processes provides its readers a unique opportunity to learn about various processes involved in meta-analysis. The first part of the book contains an introduction to meta-analysis, how to perform a systematic literature review (SLR), quality assessment of studies, extraction, and analysis of data. The second part contains information regarding meta-regression, network meta-analysis, issues, mistakes along with future perspectives. Importantly, the book integrates both general and detailed information. This book is written with the intent to allow a wide range of readers including students, researchers, and health care professionals (physicians, nurses, pharmacists, paramedical) and is also helpful for a beginner who would like to do a meta-analysis.

Acknowledgments

I would like to express my profound gratitude to Prof. Y.K. Gupta, President, AIIMS Bhopal and AIIMS Jammu; Prof. S.J.S. Flora, Ex-Director, NIPER Raebareli; Prof. S.S. Sharma, NIPER Mohali; Prof. C.R. Patil, RCPIPER, Shirpur; and Dr. Gautam Rath, Siksha 'O' Anusandhan University, Odisha, for their endless direction during my academic career. My sincere thanks to all my teachers [Kishan Sir, Vikram Sir, Ashwani Sir, Kedar Behera Sir, Rajesh Singh Sir, Kuljeet Sir, Akash Jain Sir, Vishal Sir, Dhruv Sir, Dhuwedi Sir, Prof. K.K. Pillai Sir, Prof. S. Raisuddin Sir, Prof. Nilanzan Saha Sir, Dr. Nidhi Bharal Ma'am, Dr. Fahad Haroon Sir, Dr. Biswa Mohan Padhy Sir, Prof. S.S. Gambir Sir, Prof. D. Sasmal Sir, Dr. Neelima Sharma Ma'am] for their constant guidance throughout my life. I am also grateful to my mentors, Dr. Jonathan Pillai, Sh. Parveen Garg, Prof. G.D. Gupta, Prof. R.K. Narang, Dr. Rahul Deshmukh, and Dr. Neeraj Mishra. Special thanks to all my friends Mr. Vishal Gautam, Dr. Kamal Kant, Dr. Subham Banerjee, Dr. Sanatnu Kaity, Dr. Amand Bhaskar, Dr. Saket, Dr. Abhishek Pandey, Dr. Naresh Rangra, Dr. Pragyanshu Khare, Dr. Dinesh, Dr. Avneesh Mishra, Dr. Rahul Shukla, Dr. Ashok K Datusalia, Dr. Arun Kumar, Dr. Gazal Sharma and so on whose contribution cannot be expressed in words. I would also like to thank all my colleagues at DIPSAR-DPSRU, ISF, Sun Pharmaceutical Limited Gurugram, and THSTI, for their motivation and support always. Last but not least, I would like to thank my parents (Shri Jai Kumar Sharma and Smt. Promila Devi), my younger brother (Arun Sharma), my wife (Dr. Ruchika Sharma), My Son (Master Nivaan Sharma), all my family members and dear students for their constant support throughout my life.

Contents

About the Author

Anoop Kumar currently works as an Assistant Professor in the Department of Pharmacology and Clinical Research, at Delhi Pharmaceutical Sciences & Research University (DPSRU), New Delhi, India. He has earlier served as an Assistant Professor and Head in the Department of Pharmacology and Toxicology at NIPER, Raebareli, as Associate Professor and Officiating Head in the Department of Pharmacology, ISF College of Pharmacy, Moga, Punjab, as Lecturer in Shanti Niketan College of Pharmacy, Mandi, BIRAC Research Scientist in Translational Health Science Institute (THSTI), Faridabad, India, Research Associate in Sun Pharmaceutical Limited, Gurugram, Haryana, India, QA officer in Cipla Pharmaceutical Limited Baddi, H.P. and DST INSPIRE JRF and SRF fellow at BIT Mesra, Ranchi. He has also served as an expert member of the Board of Studies of Clinical Research Programme of IKGPTU, Jalandhar, Punjab, an expert member of Drug Discovery Hackathon 2020 (DDH2020) initiated by AICTE, MHRD, Govt of India, Member Secretary of IAEC, and member of HEC.

Recently, he has been conferred with the "Notable Researcher Award 2022" in Pharmacology from DPSRU and also included as a top 2% Scientist of the World for the year 2021 in a study conducted by Stanford University, USA. He has served as a referee and guest editor for several international journals of repute. He has authored more than 100 research and review articles and 20 book chapters in international journals and publishers of repute. He is also the author of three books. His research interests are drug repurposing using computational, in vitro and in vivo techniques, meta-analysis, signal analysis in pharmacovigilance, and pharmacoeconomic studies. His lab gets funding from DST SERB, ICMR, and IITD.

Abbreviations

AI	Artificial intelligence
AMSTAR 2	Assessing the methodological quality of systematic reviews
CASP	Critical Appraisal Skills Programme
CDSR	Cochrane Database of Systematic Reviews
CEBM	Centre for Evidence-Based Medicine
CI	Confidence interval
CMA	Comprehensive meta-analysis
DARE	Database of Abstracts of Reviews of Effects
FINER	Feasible, interesting, novel, ethical, and relevant
GUI	Graphic user interface
HCPs	Health care professionals
ICTRP	Clinical Trials Registry Platform
INPLASY	The International Database to Register your Systematic Reviews
JBI checklists	Joanna Briggs Institute
MD	Mean difference
MeSH	Medical subject headings
MH	Mantel-Haenszel
MINORS	Methodological index for non-randomised studies (MINORS)
ML	Machine learning
MOOSE	Meta-analyses of observational studies in epidemiology
NDA	New drug applications
NIH	National Institutes of Health
NOS	Newcastle-Ottawa Scale
OR	Odd ratios
PICO	Patients, intervention, comparator, outcome
PRISMA	Preferred Reporting Items for Systematic Reviews and Meta-Analyses
PROSPERO	International prospective register of systematic reviews
QUOROM	Quality of reports of meta-analyses of randomised controlled trials
RCTs	Randomised controlled trials
RoB	Risk-of-bias
RR	Relative risk
SDs	Standard deviation

SEM	Standard error of mean
SLR	Systematic literature review
SMD	Standard mean difference
SND	Standard normal deviation
SPSS	Statistical package in the social sciences
STROBE	Strengthening the Reporting of Observational Studies in Epidemiology

Introduction

1

Abstract

Meta-analysis is one of the quantitative analyses that help in clinical decision-making, particularly in complex issues. The results of the individual studies that can be combined are integrated using suitable statistical procedures. The well-conducted meta-analyses provide a more precise overall estimate along with a confidence interval as well as explain heterogeneity among individual studies, the effects of outliers on the overall estimate measure, and publication bias. A systematic literature review (SLR) must be conducted as per the standard guidelines like PRISMA before conducting a meta-analysis using available software. This chapter provides a brief introduction to meta-analysis, along with an overview of the steps involved in this process.

Keywords

Randomised controlled trials (RCTs) · Systematic literature review (SLR) · Quality assessment · Models · Software · Meta-analysis

1.1 Introduction

A meta-analysis is a quantitative approach that analyses the results of already conducted studies, each addressing a closely related clinical question to derive a valid conclusion (Patel 1989). If we look at the hierarchy of evidence, meta-analyses are at the top, followed by randomised clinical trials (RCTs), whereas case reports, case series, and animal research are at the bottom, as presented in Fig. 1.1.

Generally, we prefer randomised, controlled clinical trials (RCTs) for meta-analysis; however, depending upon the objective and availability, other types of studies such as observational (case-control studies, cohort studies, cross-sectional studies, etc.) are also considered (DerSimonian and Laird 2015). The meta-analysis

A. Kumar, *Meta-analysis in Clinical Research: Principles and Procedures*,
https://doi.org/10.1007/978-981-99-2370-0_1

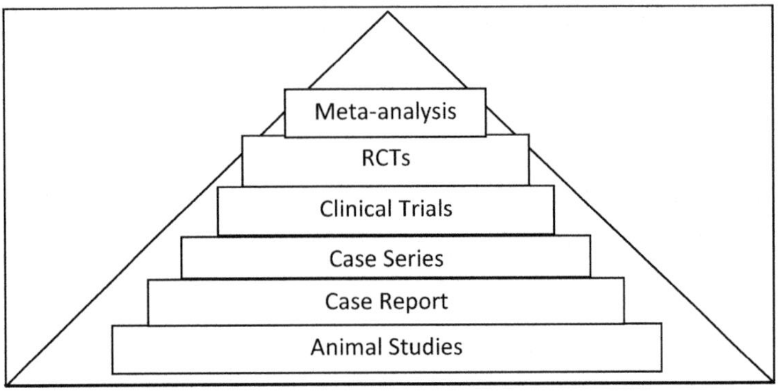

Fig. 1.1 Hierarchy of evidence

of preclinical studies (in vitro and animal studies) are also performed by researchers to draw a valid conclusion regarding particular topic. This chapter provides brief information about the meta-analysis, why is it needed? how much popular are these studies? Finally, the chapter concludes with the steps involved in this process.

1.2 Need of Meta-Analysis

The number of clinical studies being published on a daily basis is increasing significantly. Practically, it is very difficult for physicians to process large amounts of new information available from these published studies, particularly to resolve conflicting findings. Furthermore, the results of individual studies are not consistently reproducible, thus, not enough to provide confidence. Usually, most studies are conducted by different research teams across the globe and sometimes, the results of these studies particularly studies with small sample sizes are conflicting and ultimately arise confusion among physicians to make better clinical decisions. Thus, there is a need for suitable approaches that help the physician in clinical decision-making. A meta-analysis helps to resolve conflict among studies using statistical approaches.

The efficacy of drugs is tested in clinical trials, most usually in Phase 2 and Phase 3 clinical trials, and based on the data generated from these trials, sponsors submit new drug applications (NDA) to regulatory authorities to get market approval. However, the sample size in clinical trials is too small to detect the effects of drugs in a real-world population. Therefore, we found a number of trials with a small sample size, when we search for studies to answer clinical questions (Gøtzsche 2000). Therefore, meta-analysis will help to combine the results of individual studies to derive a valid conclusion.

1.3 Popularity of Meta-Analysis

I have just searched for the keyword 'meta-analysis' in PubMed and found over 150,140 papers so far, while 'meta-analysis and COVID-19' alone accounted for over 1730 articles in the last year which indicates the popularity of meta-analyses. I have also found some articles that performed a meta-analysis of meta-analyses, so-called meta-meta-analysis. Figure 1.2 depicts the year-by-year popularity of meta-analysis based on data from one of the search engines, PubMed.

1.4 Guidelines and Checklists

It is always recommended for researchers to follow a specific guideline to conduct a meta-analysis. The Preferred Reporting Items for Systematic Reviews and Meta-Analyses (PRISMA) guideline is the most well-known (Sarkis-Onofre et al. 2021). The various checklists are also available to ensure the quality and reliability of the results of the meta-analysis. A few examples are the quality of reports of meta-analyses of randomised controlled trials (QUOROM) (Moher et al. 2000), the MOOSE (meta-analyses of observational studies in epidemiology) Checklist (Brooke et al. 2021), study quality assessment tools like the NIH quality assessment tools for case-control studies, controlled intervention studies, cohort studies (Zeng et al. 2015), the Newcastle-Ottawa Scale (NOS) (Lo et al. 2014), and so on. The details regarding commonly used scales for quality assessment of studies are explained in Chap. 3.

1.5 Steps Involved

Various steps are involved to perform a meta-analysis. Typically, the following main steps are involved to perform a meta-analysis.

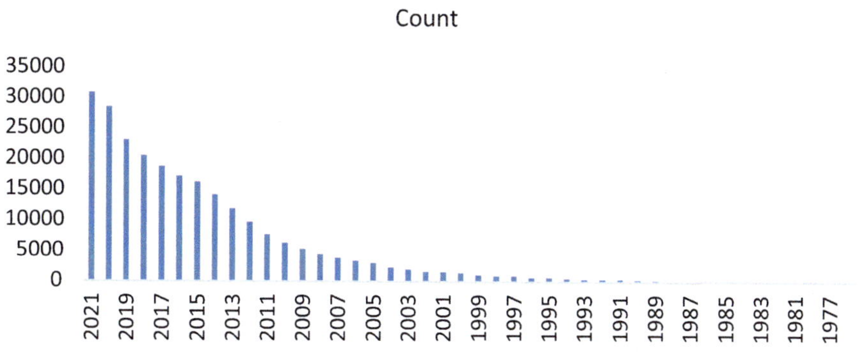

Fig. 1.2 Popularity of meta-analysis (till Dec 2021)-extracted from PubMed

1. State the problem
2. Define your research questions
3. Search strategy
4. Sorting of studies
5. Quality assessment
6. Extraction of data
7. Analysis of data
8. Interpretation of data
9. Reporting of data

The first step is to clearly define the problem. For example: there is confusion among physicians regarding the use of steroids in COVID-19 patients. Thus, in this particular example, researchers should define the research problem clearly, particularly the outcome. Second step: convert your problem into a PICO format which means which type of patients (P)? What is intervention (I)? What about control or comparator (C)? What will be the outcome (O)? Thus, in this particular example, patients: COVID-19, Intervention: steroids, Comparator: without steroids, Outcome: Death. The third step is to search the available databases like PubMed, Medline, clinical trial websites, etc. with the proper Medical Subject Headings terms. The fourth step is the sorting of studies as per inclusion and exclusion criteria (depending upon the objective of the study). The detail about the search in available databases is explained in Chap. 2. The fifth step is to perform a quality assessment of selected studies using standardised scales like Newcastle Ottawa, National Institute of Health (NIH) scale, etc. as the quality of a study can affect the results. The details about the quality assessment of studies are mentioned in Chap. 3. The next step is the collection of data from selected studies based on quality assessment. The data should be collected in a validated form which should contain all the important columns as per the design of the study. I personally feel most of the errors occurred during the extraction of the data. Therefore, I always recommend authors to extract the data in validated form, at least by two independent authors. The next step is the analysis of the data by calculating the overall effect by combining the data from the individual studies using suitable statistical procedures. Suppose we would like to know the effect of a particular intervention on the blood pressure of patients. In this case, endpoint is continuous, so the author should calculate an overall estimate of all individual studies in terms of the mean difference. If our data is categorical, suppose we would like to know the effect of a particular intervention on the deaths of patients, in this particular case, the overall estimate should be calculated as an odds ratio or relative risk instead of the mean difference. The extraction and analysis of data are explained in Chap. 4. The selection of a model is also very important when performing a meta-analysis. The fixed effect and random effect models are the most commonly used models. The details about the selection of a model are explained in Chap. 5. As results of individual studies are combined together, thus heterogeneity among studies is very important to calculate. Another important aspect is the determination of publication bias qualitatively (funnel plot) and quantitatively (Begg and Egger tests). The details regarding heterogeneity among studies and publication bias are explained in Chap. 6. It is very common to get outliers, when we are combining the results of individual studies. Thus, sensitivity analysis is also important to check the effects of outliers on the outcome. The subgroup analysis should also be performed to check the effect of individual variables on the outcome depending upon the availability of the studies. The details about sensitivity and subgroup analysis are given in Chap. 7. The last step is to present the results of the meta-analysis graphically. The most common plots in the meta-analysis are forest and funnel plots. The forest plot represents the individual study estimate along with a 95% confidence interval as well as pooled estimate of all individual studies, whereas the funnel plot represents the qualitative analysis of publication bias. Various plots in meta-analysis are presented along with suitable examples in Chap. 9 (Field and Gillett 2010; Borenstein et al. 2021; Hedges and Olkin 2014; Egger et al. 2002; Hartung et al. 2008; Higgins and Thompson 2002).

1.6 Available Software

Various software are available to perform the meta-analysis. The most commonly used software are REVMAN, STATA, EPIMETA, meta R package in R, etc.

1.7 Conclusion

Meta-analyses are applicable not only to the medical field but also to other fields. It is a powerful tool to combine the results of individual studies to derive valid conclusions particularly when individual studies are inconclusive. However, these studies should be planned and executed carefully as the number of factors could affect the results of these studies. It is always recommended to involve a group of experts to conduct this type of analysis in a proper way.

References

Borenstein M, Hedges LV, Higgins JP, Rothstein HR (2021) Introduction to meta-analysis. Wiley, New York

Brooke BS, Schwartz TA, Pawlik TM (2021) MOOSE reporting guidelines for meta-analyses of observational studies. JAMA Surg 156(8):787–788

DerSimonian R, Laird N (2015) Meta-analysis in clinical trials revisited. Contemp Clin Trials 45: 139–145

Egger M, Ebrahim S, Smith GD (2002) Where now for meta-analysis? Int J Epidemiol 31(1):1–5

Field AP, Gillett R (2010) How to do a meta-analysis. Br J Math Stat Psychol 63(3):665–694

Gøtzsche PC (2000) Why we need a broad perspective on meta-analysis: it may be crucially important for patients. BMJ 321(7261):585–586

Hartung J, Knapp G, Sinha BK (2008) Statistical meta-analysis with applications, vol 6. Wiley, New York

Hedges LV, Olkin I (2014) Statistical methods for meta-analysis. Academic Press

Higgins JP, Thompson SG (2002) Quantifying heterogeneity in a meta-analysis. Stat Med 21(11):1539–1558

Lo CKL, Mertz D, Loeb M (2014) Newcastle-Ottawa scale: comparing reviewers' to authors' assessments. BMC Med Res Methodol 14(1):1–5

Moher D, Cook DJ, Eastwood S et al (2000) Improving the quality of reports of meta-analyses of randomized controlled trials: the QUOROM statement. Rev Esp Salud Publica 74(2):107–118

Patel MS (1989) An introduction to meta-analysis. Health Policy 11:79–85

Sarkis-Onofre R, Catalá-López F, Aromataris E, Lockwood C (2021) How to properly use the PRISMA statement. Syst Rev 10(1):1–3

Zeng X, Zhang Y, Kwong JS (2015) The methodological quality assessment tools for preclinical and clinical studies, systematic review and meta-analysis, and clinical practice guideline: a systematic review. J Evid Based Med 8(1):2–10

Systematic Literature Review (SLR) 2

Abstract

Systematic literature review (SLR) is the first and most important step to perform any kind of meta-analysis. However, every systematic review is not possible to convert into a meta-analysis, but every meta-analysis starts with a systematic review only. SLR is designed to answer a specific research question by performing a systematic analysis of literature as per the standard guidelines like PRISMA. This chapter provides a detailed discussion on SLR.

Keywords

Systematic review · PRISMA guidelines · MeSH terms

2.1 Introduction

The medical literature is growing very fast, and it is not possible for health care professionals (HCPs) to assess this vast variety of research. Systematic literature reviews (SLRs) are used to provide up-to-date knowledge of particular issues on intervention, diagnostic tests, health, etc. in a systematic way to provide answers to specific research questions. The main purpose of a systematic review is to present the current state of knowledge regarding a particular issue and to fill the important gaps in knowledge. The review can be updated from time to time depending upon the availability of studies regarding a particular issue. The systematic review can only be done if the aim and objectives are clear to the researchers.

The methodology for systematic review should be structured, transparent, and reproducible, which will provide reliable findings to make a valid conclusion. Briefly, a researcher should be clear about scope, eligibility criteria (inclusion and exclusion), searching of all possible databases, issues of bias, and analysis of included studies (Baker and Weeks 2014; Xiao and Watson 2019; Whitea and

A. Kumar, *Meta-analysis in Clinical Research: Principles and Procedures*, https://doi.org/10.1007/978-981-99-2370-0_2

Schmidtb 2015). Of course, the development of a good SLR is a time-consuming process, and expertise is required. This chapter provides detailed information regarding the systematic literature review (SLR) along with the steps involved in this process.

2.2 Important Points

There are some important points for researchers who would like to do SLR. Researchers should involve a team to perform SLR. Usually, SLR done by a single author is not considered by most of the journals. The purpose of involving a team is to reduce errors by performing the same tasks like the selection of studies, extraction of data, and quality assessment by at least two researchers independently. It is always recommended for researchers who are working for the first time in SLR to involve at least one person in their team who has previous experience in SLR as well as expertise in that particular topic.

Statistical expertise is also required if researchers are planning for a meta-analysis. It is better to include different experts, such as methodological experts, topic experts, and analysis experts in your team to perform SLR effectively. The scope of review should also be clear to the researchers to avoid the risk of conceptual incoherence. It is also very important to cross-check already published literature as well as ongoing or completed reviews in registry databases of SLR such as PROSPERO before starting a review (PROSPERO 2021). It will help in the reduction of duplicates. The outcome of the review should address important issues that are relevant to consumers, health professionals, and policymakers.

2.3 Importance of SLR

SLRs are important to address important research questions in a systematic way. The scientific gap in a particular field as well as priorities for further research are also identified using SLR. There is a great demand for high-quality SLR; however, it is a time-consuming process.

2.4 Difference Between Narrative and Systematic Literature Review (SLR)

A review is generally synthesising the main findings of two or more publications on a given topic, whereas a systematic review identifies and tracks down all the literature on a given topic. For pooling the results of individual studies into a single estimate, the meta-analysis employs a specific statistical strategy (Paul and Leibovici 2014; Tawfik et al. 2019).

All meta-analysis is a systematic review, but all systematic reviews are not a meta-analysis. It depends upon the availability of data specifically regarding the aim

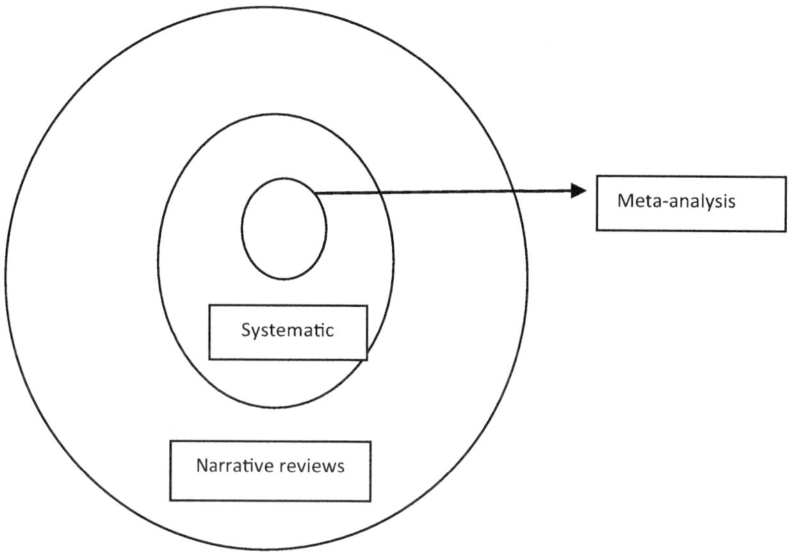

Fig. 2.1 Graphical representation of the difference between narrative review, systematic review, and meta-analysis

Table 2.1 Main differences between narrative, systematic review, and meta-analysis

Feature	Narrative review	Systematic review	Meta-analysis
Methodology	No	Yes	Yes
Extensive searching of databases and clinical trial websites	No	Yes	Yes
Quality assessment	No	Yes	Yes
Weightage to studies	No	Yes	Yes
Assessment of bias	No	Yes	Yes
Statistical methods	No	No	Yes

and objective of the review. Figure 2.1 shows the differences between narrative, systematic review, and meta-analysis (Table 2.1).

2.5 Protocol Development

It is always recommended to develop a well-defined protocol before starting an SLR. The protocol should clearly describe objectives, a systematic approach for eligibility criteria, searching strategies, quality assessment of studies, methods, analysis, and representation of results. The protocol should describe the plans for the minimization of bias. Usually, the biases in clinical studies are raised by various sources, such as the randomization process of participants, selection of particular participants, inappropriate blinding of study participants, inappropriate collection and analysis of data. An author should also plan at beginning of SLR regarding the selection of overall estimate measures like mean difference, standard mean difference, odds ratio, risk ratio, risk difference, etc., assessment of heterogeneity, publication bias, and selection of a model (random or fixed effect model). The author should also plan how to address heterogeneity among studies? Which factors could affect the results of the study?

Generally, the SLR should adhere to the designed protocol and the author should make every effort to adhere to the designed protocol. However, sometimes changes in protocol are required. This is also the case for a well-designed randomised controlled trial (RCT), where sometimes changes in protocol are required due to some unavoidable problems such as lower recruitment of subjects, unexpected results, and so on. However, it is very important that changes should not be made based on the outcome of the research. The changes in inclusion and exclusion criteria as well as changes in statistical analysis after completion of a study are highly susceptible to bias, and therefore, should be avoided unless there is a strong justification for doing this.

Overall, every SLR must have a predesigned protocol and I always recommend authors to register their designed protocol with a standard register of SLR like CDSR, PROSPERO, etc., which helps in the reduction of duplicates, promotes accountability, and saves effort and time of authors. The well-designed protocol ensures that SLR is most likely to help physicians, policymakers, government agencies, and consumers to make better clinical decisions. The changes should be mentioned clearly if done (Linares-Espinós et al. 2018).

2.6 Steps to Perform SLR

2.6.1 Frame Your Objective and Research Questions

The objective of the review should be clear, which helps to set the eligibility criteria. The research questions can be broad or narrow depending upon the aim of the review. For example: the review might address the role of steroids in the reduction of mortality of serious COVID-19 patients or can address particular steroid-like the role of dexamethasone role in the reduction of mortality of serious COVID-19 patients. Generally, reviews are conducted in a broader way, followed by narrow questions depending on the availability of data from studies. If research questions are related to the effects of particular interventions, the PICO format, which is an acronym for Population (P), Intervention (I), Comparison(s) (C), and Outcome (O), is usually preferred. The author should define the populations. For example: researchers would like to review the role of steroids in COVID-19 patients. So, the author should clearly define the type of patients they are dealing with (mild, moderate, severe, patients with co-morbid conditions)? Next is intervention, which should also be clearly defined. In this particular example, if researchers would like to review the role of steroids, which types of steroids? At what dosage? Which route of administration? What dosage regimen? Will studies include the combination of steroids with other classes of drugs? Next, the author should define the comparator. In this particular example, Will the comparison be between the steroidal and non-steroidal groups or the steroidal with the control group? However, for every SLR, each PICO component is not necessary, it depends upon the research questions. For example, if researchers would like to find out a set of biomarkers for the identification of a particular disease. In this particular case, intervention is not applicable.

It is better if the author could frame their research questions as per the FINER criteria which means the questions should be Feasible, Interesting, Novel, Ethical, and Relevant (Farrugia et al. 2010).

2.6.2 Define Eligibility Criteria

Practically, it is not possible to include all the studies that are identified after an extensive search. Therefore, studies should be screened as per the inclusion and exclusion criteria. The next step is to define your inclusion and exclusion criteria as per the objective of your review. The authors should decide whether they will include only published studies or unpublished studies also as the reliability of unpublished studies is unclear. The search engines like PubMed, Medline, etc. are most commonly used to search published studies. However, using only electronic databases could miss important studies. Therefore, the author should search other databases also, like the Cochrane Controlled Trials Register which is one of the best electronic sources to search for clinical trials, clinical trial registries, etc. The bibliographies of already published review articles, monographs, etc. should also be scrutinised. The authors should also plan in advance regarding methods for the identification of studies as per the eligibility criteria. Practically, the author is not able to get some relevant studies due to the language of the publication, which could result in publication bias. Therefore, it is always recommended to design searches in such a way that many eligible studies can be captured from all possible sources to reduce publication bias.

The author should also define which types of studies will be included? Normally, RCTs are included. However, non-randomised studies are also included, particularly when randomization is not possible or randomised trials are unable to address the effect of a particular drug in a specific type of population.

Some studies could contain errors. Thus, it is very important to examine studies carefully for any retraction statements or errata since publication. Normally, errata are published to correct errors that do not affect the scientific content of the article. Therefore, these studies can be included after discussion with all the authors. However, retractions are those which affect the scientific content of the article. Therefore, data from these studies should not be included in SLR as it will impact the overall estimate measure in a significant way.

2.6.3 Search Strategy

After defining the eligibility criteria, the next step is to search the available databases like PubMed, Medline, and Embase with the proper MeSH terms. The most popular databases are PubMed/Medline (http://www.ncbi.nlm.nih.gov), Embase (http://www.elsevier.com/online-tools/embase), Cochrane Library (http://www.cochranelibrary.com). The other search strategies are hand search (from the library), references from retrieved articles. The clinical trials are usually registered with trial registries like the Clinical Trial Registry of India (www.ctri.in), Global Clinical Trials at www.clinicaltrials.gov.in, the WHO International Clinical Trials Registry Platform (ICTRP) portal, etc. The author should also contact particular researchers working in particular areas of interest as well as organisations for information about unpublished or ongoing studies. The author should also search for reports, thesis, and abstracts of relevant conferences.

The search of electronic engines should be done with proper Medical Subject Headings (MeSH), and Boolean operators 'AND' and 'OR' should be used in the correct manner. In every topic, the terms are joined together with the Boolean 'OR' operator, and the concepts are combined with the Boolean 'AND' operator. There is another operator 'NOT' which should be avoided as much as possible as it could arise the fear of removing studies that might be relevant. The author could also use search filters as per the objective of the study. The search filters will help the author to retrieve specific types of studies. For example, the author would like to include only Randomized controlled trials (RCTs), so a filter can be applied that will help to identify only RCTs. Many errors have been observed in search strategies or search strategies are not framed in a standard way, which results in poor quality of SLR. The search strategies are one of the important steps in the execution of high-quality SLR and should be peer-reviewed. I always recommend my students to show me search strategies before running the searches. The authors are also encouraged to use standard checklists like the PRESS Evidence-Based Checklist to assess electronic search strategies. These checklists are helpful to include all relevant aspects of the designed protocol and ultimately improve the quality of searches.

2.6.4 Sorting of Studies

The next step is the sorting of studies. Duplicates should be removed first, but the identification of duplicates needs some detective work. The duplicates can be identified through the registration number of clinical trials, name of authors, location of authors, specific details about drugs, number of participants, baseline data, and duration of the study. Software such as reference management software is helpful in the identification and removal of duplicates. The next step is the sorting of studies based on titles. The author should read the titles carefully and irrelevant titles should be excluded with valid reasons. Furthermore, the abstracts of relevant studies should be extracted and screened as per the inclusion and exclusion criteria of the study. The irrelevant abstracts should be excluded with proper justification. Finally, the full text of relevant abstracts should be retrieved and screened as per the eligibility criteria. I always assign the SLR to at least two groups of students on the same topic and compare the eligible studies. The conflict among studies is discussed among all the authors and they all take an appropriate decision. The author could also check with the corresponding author of the published study to sort out confusion if any. The author could also request the corresponding author to provide more detailed information regarding methods, results, etc. if required. It is also recommended that incomplete studies or studies that have not been obtained should be compiled as a separate table of studies with proper labels.

The decision regarding inclusion and exclusion criteria is the most important part of any SLR and usually involves judgement. It is always recommended to involve at least two people to work independently in the screening of studies as per the inclusion and exclusion criteria of the study. The disagreements between the authors should be resolved through discussion. However, if disagreements are still not resolved, the third author should be consulted.

The selection process should be documented in a proper flow diagram, as well as tables should be created regarding the characteristics of included as well as excluded studies. The valid reason for the exclusion of studies should be documented. The reason for the exclusion of a particular type of studies need not be documented. For example, the author would like to perform the SLR of RCTs only. So, in this case, the reason for the exclusion of individual non-randomised studies need not be documented. The step-by-step details of inclusion, as well as exclusion of studies, should be presented in a flow diagram as per the standard guidelines like PRISMA (Page et al. 2021). The flow diagram should be completed correctly (Muka et al. 2020).

2.6.5 Quality Assessment

A quality assessment of studies should be done. Various checklists, such as Downs and Black (Downs and Black 1998), NIH quality assessment scale, Cochrane risk-of-bias (RoB 2) tool, Joanna Briggs Institute (JBI) checklists, etc., are available for quality assessment of studies. The details of various commonly used checklists for quality assessment of studies are discussed in Chap. 3.

2.6.6 Collection of Data

The relevant data from full-text studies should be extracted in a predesigned form as per the objective of the study. Suppose an author would like to extract the data related to deaths of COVID-19 patients in the aspirin group as compared to the non-aspirin group. So, the outline of the tables will include the following headings: name of the author, total sample size, location of study, number of males/females, number of subjects in the aspirin group, number of subjects who died in the aspirin group, number of subjects in the non-aspirin group, number of subjects died in the non-aspirin group.

The author should design easy-to-use forms that should contain all relevant headings as per the objective of the study. The author should collect data in such a way that meta-analysis can be done if possible.

2.6.7 Analysis

The risk of bias in the included studies should be assessed and can be categorised into low risk, high risk, and unclear risk of bias.

2.7 Conclusion

Systematic literature reviews help clinical decision-makers to access quality information on particular topics that is up-to-date and systematic. The SLR should always be conducted by a team that should include individuals with different expertise. An author should keep all the records related to SLR, starting from the inclusion or exclusion of studies to the compilation of data. Good data management is essential to complete SLR in a proper way. Future efforts should also be made to update the SLR from time to time.

References

Baker KA, Weeks SM (2014) An overview of systematic review. J Perianesth Nurs 29(6):454–458

Downs SH, Black N (1998) The feasibility of creating a checklist for the assessment of the methodological quality both of randomised and non-randomised studies of health care interventions. J Epidemiol Community Health 52(6):377–384

Farrugia P, Petrisor BA, Farrokhyar F et al (2010) Practical tips for surgical research: research questions, hypotheses and objectives. Can J Surg 53(4):278–281

Linares-Espinós E, Hernández V, Domínguez-Escrig JL et al (2018) Methodology of a systematic review. Actas Urol Esp (Engl Ed) 42(8):499–506

Muka T, Glisic M, Milic J et al (2020) A 24-step guide on how to design, conduct, and successfully publish a systematic review and meta-analysis in medical research. Eur J Epidemiol 35(1): 49–60

Page MJ, McKenzie JE, Bossuyt PM et al (2021) The PRISMA 2020 statement: an updated guideline for reporting systematic reviews. J Clin Epidemiol 134:178–189

Paul M, Leibovici L (2014) Systematic review or meta-analysis? Their place in the evidence hierarchy. Clin Microbiol Infect 20(2):97–100

PROSPERO (2021) International prospective register of systematic reviews. https://www.crd.york.ac.uk/prospero/. Accessed 10 Nov 2021

Tawfik GM, Dila KAS, Mohamed MYF (2019) A step by step guide for conducting a systematic review and meta-analysis with simulation data. Trop Med Health 47(1):1–9

White A, Schmidt K (2015) Systematic literature reviews. Complement Ther Med 13:54–60

Xiao Y, Watson M (2019) Guidance on conducting a systematic literature review. J Plan Educ Res 39(1):93–112

3 Quality Assessment of Studies

Abstract

The quality assessment of studies is one of the important steps in meta-analysis. Various scales and checklists are available to assess the quality of studies, depending upon the type of study. A detailed discussion regarding the assessment of the quality of studies using available scales/checklists is compiled in this chapter.

Keywords

Quality assessment · Scales · Checklists

3.1 Introduction

The assessment of the quality of studies is very important as it will directly impact the results of SLR. The results of poor-quality studies are not reliable as many factors could influence the conclusion of the study. Therefore, these studies should be excluded from the analysis. Currently, various tools like Newcastle Ottawa, NIH scale, etc. are available depending upon the type of studies (case-control studies, cohorts, RCTs, etc.) for the quality assessment of studies. Some checklists are built into software which is frequently used for quality assessment of studies. Ideally, authors should consider only randomised controlled trials (RCTs) to draw a valid conclusion. However, in some cases, depending upon the aim and objective of the review, other types of studies of good quality are also included. This chapter discusses the most commonly used tools for the quality assessment of studies.

A. Kumar, *Meta-analysis in Clinical Research: Principles and Procedures*,
https://doi.org/10.1007/978-981-99-2370-0_3

3.2 Checklists/Scales

Various checklists or scales are available for the quality assessment of studies. A few of them are mentioned below.

1. Assessing the Methodological Quality of Systematic Reviews (AMSTAR 2)
2. Centre for Evidence-Based Medicine (CEBM) critical appraisal tools
3. Cochrane risk-of-bias (RoB 2) tool
4. Critical Appraisal Skills Programme (CASP)
5. Joanna Briggs Institute (JBI) checklists
6. Newcastle-Ottawa Scale (NOS)
7. Quality Assessment of Diagnostic Accuracy Studies (QUADAS-2) tool
8. Strengthening the Reporting of Observational Studies in Epidemiology (STROBE) framework
9. Jadad scale
10. van Tulder scale
11. CCRBT
12. GRADE
13. AHRQ
14. Database of Abstracts of Reviews of Effects (DARE)
15. Downs and Black Checklist
16. GRACE checklists
17. Methodological Index for Non-randomised studies (MINORS)

These checklists or scales are discussed in detail in the following sections.

3.2.1 Assessing the Methodological Quality of Systematic Reviews (AMSTAR 2)

A MeaSurement Tool to Assess systematic Reviews (AMSTAR) is designed to appraise systematic reviews that contain RCTs. AMSTAR was published in 2007 and revised and updated depending upon the feedback and critiques of researchers. AMSTAR-1 is replaced with AMSTAR-II. The main changes include simplification of response categories, detailed consideration of the Risk of bias (Rob), better alignment with the PICO framework, and more detailed information on inclusion and exclusion studies. The AMSTAR is available at www.amstar.ca (Shea et al. 2017).

3.2.2 Risk of Bias in Systematic Reviews (ROBIS)

Another instrument, the Risk Of Bias In Systematic reviews (ROBIS) is a comprehensive tool for non-randomised studies, released in 2016. This tool is mainly focused on the risk of bias, containing three phases that cover most types of research questions (Whiting et al. 2016).

3.2.3 Centre for Evidence-Based Medicine (CEBM)

The Centre for Evidence-Based Medicine (CEBM) contains a collection of critical appraisal tools to appraise the reliability, importance, and applicability of clinical evidence. These tools contain a variety of questions such as: Does this study address a clearly focused question? Did the study use valid methods to address this question? Are the valid results of this study important? Are these valid, important results applicable to patients or population? Depending upon the type of medical evidence, appraisal sheets are available for the evaluations, such as the Randomised Controlled Trials (RCT) Critical Appraisal Sheet, Diagnostics Critical Appraisal Sheet, Systematic Reviews Critical Appraisal Sheet, Prognosis Critical Appraisal Sheet, Critical Appraisal of Qualitative Studies Sheet, and IPD Review Sheet. These appraisal sheets are available in various languages such as Chinese, German, Lithuanian, Portuguese, and Spanish (Centre for Evidence-Based Medicine (CEBM) tools 2022).

3.2.4 Cochrane Risk-of-Bias (RoB 2) Tool

The Cochrane risk-of-bias (version 1) tool has been revised as RoB 2 (version 2). The tool's second version replaces the first, which was first published in the Cochrane Handbook's Version 5 in 2008 and revised in 2011. This tool is mainly recommended for the assessment of the risk of bias in RCTs. RoB 2 is divided into a collection of bias domains that focus on various aspects of trial design, conduct, and reporting. A set of questions are asked within each domain to extract information on aspects of the trial that are important to the risk of bias. Based on the answers to the set of questions, an algorithm generates a proposed judgement regarding the risk of bias resulting from each area. The risk of bias in a judgement can be 'Low' or 'High,' or it can indicate 'Some worries' (Jørgensen et al. 2016). The field of research has progressed, and RoB 2 represents the current understanding of how bias reasons might affect study results, as well as the best strategies to detect this risk. The tool is available now, but it will only be deployed in the Review Manager software 2020 online edition (Review Manager (RevMan) 2020).

3.2.5 Critical Appraisal Skills Programme (CASP)

Critical appraisal abilities allow us to evaluate the credibility, relevance, and outcomes of published articles in a methodical way. The Critical Appraisals Skills Programme (CASP) has over 25 years of unrivalled experience in providing healthcare professionals with training. This collection of eight critical appraisal techniques is intended for use while reading research. Systematic Reviews, Randomised Controlled Trials, Cohort Studies, Case-Control Studies, Economic Evaluations, Diagnostic Studies, Qualitative Studies, and Clinical Prediction Rule are all covered by CASP's appraisal checklists (Critical Appraisal Skills Programme (CASP) Checklists 2022).

Checklist for CASP Randomised Controlled Trials
Checklist for CASP Systematic Review
Checklist for CASP Qualitative Studies
Checklist for CASP Cohort Studies
Checklist for CASP Diagnostic Studies
Checklist for CASP Case-Control Studies
Checklist for Economic Evaluation via CASP
Checklist of CASP Clinical Prediction Rules

3.2.6 Joanna Briggs Institute (JBI) Checklists

The critical appraisal techniques developed by JBI aid in determining the reliability, relevance, and findings of published works (Joanna Briggs Institute (JBI) 2022). The various checklists provided by JBI are mentioned below.

Analytical Cross-Sectional Studies Checklist
Case-Control Studies Checklist
Case Reports Checklist
Case Series Checklist
Cohort Studies Checklist
Diagnostic Test Accuracy Studies Checklist
Economic Evaluation Checklist
Prevalence Studies Checklist
Qualitative Research Checklist
Quasi-Experimental Studies Checklist
Randomised Controlled Trials Checklist
Systematic Reviews Checklist
Text and Opinion Checklist

3.2.7 Newcastle-Ottawa Scale (NOS)

The quality of the included studies particularly observational studies can also be assessed using the Newcastle-Ottawa scale (NOS). NOS is based on a star rating system, with each study receiving a maximum of nine (for prospective and cross-sectional studies) and 10 (for case-control studies) stars. Two authors separately checked the quality of the work, and any conflicts were resolved by the third. Studies with a score of 6 or higher were regarded as being of excellent quality (Margulis et al. 2014).

3.2.8 Quality Assessment of Diagnostic Accuracy Studies (QUADAS-2) Tool

The QUADAS tool was created in 2003 to conduct systematic evaluations of diagnostic accuracy studies. QUADAS-2 was created as a result of experience, anecdotal reports, and feedback indicating areas for improvement. Patient selection, index test, reference standard, and flow, as well as timing, are the four areas of this tool. Each domain is evaluated for bias risk, with the first three domains also being evaluated for applicability concerns. Signalling questions are added to aid in determining the likelihood of bias. The QUADAS-2 tool is used in four stages: summarising the review question, tailoring the tool and producing review-specific instructions, creating a flow diagram for the primary study, and judging bias and applicability. This tool makes the bias and applicability of primary diagnostic accuracy studies more evident (Whiting et al. 2011).

3.2.9 Strengthening the Reporting of Observational Studies in Epidemiology (STROBE)

STROBE stands for strengthening the reporting of observational studies in epidemiology, an international collaborative endeavour comprising epidemiologists, methodologists, statisticians, researchers, and journal editors involved in the conduct and dissemination of observational studies.

The assessment of the strengths and shortcomings of studies reported in the medical literature is hampered by incomplete and insufficient study reporting. Readers must understand what was planned (and what was not), what was done, what was discovered, and what the outcomes mean. Recommendations from top medical publications on how to report studies can help to enhance the quality of reporting.

Observational research covers a wide range of study designs and topics. The STROBE Statement was created to offer a checklist of topics that should be included in articles reporting such research. The most common observational studies are cohort, case-control, and cross-sectional studies (Skrivankova et al. 2021; von Elm et al. 2007). The STROBE Checklists are available for these studies. The STROBE Initiative's documentation and publications are all open access and available for download on this website https://www.strobe-statement.org/ (Strengthening the reporting of observational studies in epidemiology 2022).

3.2.10 Jadad Scale

The Jadad scale, often known as Jadad scoring or the Oxford quality score system, is a mechanism for evaluating a clinical trial's methodological quality independently. It was named after Colombian physician Alex Jadad, who established a system for assigning a score between zero (extremely poor) and five (excellent) to such trials in 1996. A Jadad score is based on a three-point questionnaire. Answering each question with a yes or no is required. Each yes would be worth one point, and each no would be worth zero; there would be no fractional points. The Jadad team estimated that scoring each paper would take no more than 10 min. The following were the questions: Was there any mention of the study being randomised? Was the research described as a double-blind study? Was there any information about withdrawals or dropouts? An article should describe the number of withdrawals and dropouts in each of the study groups, as well as the underlying causes, to obtain the appropriate point. Additional points were awarded if the randomisation process was explained in the publication and was appropriate or the blinding approach was detailed, and it was found to be appropriate. If the method of randomisation was explained but was improper, or if the method of blinding was mentioned but was inappropriate, points would be deducted. As a result, a clinical trial's Jadad score could range from zero to five. Although there are only three questions, the Jadad scale is commonly referred to as a five-point scale. It consists of five points in total: two for randomisation, two for blinding, and one for the dropout rate. One point is given in each area when the report just contains general comments and no specific description of randomisation and blinding. When the suitable procedure is described in detail, one point is added. When the technique of description is insufficient, one point is subtracted. One point is awarded when the specified number and reasons for dropouts by topic group are presented. Even if no dropouts occur, this should be noted in the statement. When the sum is three points, it is considered good quality; when the total is two points, it is considered low quality. However, if the study's design could not be double-blinded, it is considered good quality if the overall score is less than two points (Jadad et al. 1996).

3.2.11 van Tulder Scale

Randomisation, allocation concealment, baseline characteristics, patient blinding, caregiver blinding, observer blinding, co-intervention, compliance, dropout rate, endpoint assessment time point, and intention-to-treat analysis are all included in the van Tulder scale. Its evaluation approach is to select 'yes, 'no', or 'don't know' for each item, and when 5 things are satisfied (5 points), the report is considered good quality (van Tulder et al. 2003).

3.2.12 CCRBT

The CCRBT classifies RCTs into six categories: sequence creation, allocation concealment, blinding, inadequate outcome data, selective outcome reporting, and other potential validity risks. For each domain, the assessment suggests 'yes, no, or unclear,' indicating a low, high, or unclear risk of bias, respectively (Chung et al. 2013).

It is categorised as having a low risk of bias when the first three questions are answered with 'yes' and no significant issues about the last three domains are detected, while it is classified as having a moderate risk of bias when it is assessed in two domains with 'unclear' or 'no'. The cases that scored 'unclear' or 'no' in three domains were classified as having a high risk of bias.

3.2.13 GRADE

The GRADE (short for Grading of Recommendations Assessment, Development, and Evaluation) working group began in the year 2000 as an informal collaboration of people interested in addressing the flaws in health-care grading systems. The working group has created a uniform, logical, and transparent approach to rating evidence quality (or certainty) and recommendation strength. Many worldwide organisations contributed to the development of the GRADE approach, which is currently the industry standard for developing guidelines (Atkins et al. 2004).

3.2.14 Avoiding Bias in Selecting Studies (AHRQ)

Systematic reviews include a risk-of-bias evaluation, although there is little definitive empirical evidence on the validity of such assessments. In the face of such uncertainty, the Evidence-based Practice Centres created practical suggestions that can be used consistently across review subjects, enhance transparency and reproducibility in procedures, and address methodological advances in risk-of-bias assessment (Agency for Healthcare Research and Quality 2022).

3.2.15 Database of Abstracts of Reviews of Effects (DARE)

The Centre for Reviews and Dissemination (CRD) is an international organisation dedicated only to the synthesis of evidence in the field of health. The CRD also maintains a number of databases that are widely used by health professionals, policymakers, and researchers worldwide. The CRD also conducts methodological research and publishes internationally recognised standards for conducting systematic reviews.

DARE (Database of Abstracts of Reviews of Effects): CRD researchers conduct a systematic search of the world literature to discover and describe systematic reviews, as well as assess their quality and identify their strengths and limitations. On a weekly basis, hundreds of citations are analysed to find possible systematic reviews using extensive search algorithms.

Two researchers independently examine those citations identified as possible systematic reviews for inclusion using the following criteria:

1. Were any criteria for inclusion or exclusion mentioned?
2. Was the search thorough enough?
3. Were the research included in the review synthesised?
4. Was the quality of the studies included in the review evaluated?
5. Are there enough facts regarding the various studies that have been included?

DARE is available free of charge on the internet http://www.crd.york.ac.uk/crdweb/ (Database of Abstracts of Reviews of Effects (DARE) 1995).

3.2.16 Downs and Black Checklist

For public health practitioners, policymakers, and decision-makers, this checklist provides a complete critical appraisal tool. The tool is a simple, step-by-step assessment that may be used with any quantitative study methodology. This checklist has 27 criteria that span a wide range of topics, including reporting quality, external and internal validity, and power. Study quality, external validity, study bias, confounding and selection bias, and study power are the five major sections. Instructions and a scoring system for each portion, as well as a final total score, are included in each segment. As a result, instead of 5, the maximum score for item 27 was 1 (based on a power analysis), and the highest possible score for the checklist was 28 (instead of 32).

Downs and Black score ranges were assigned the following quality levels: outstanding (26–28), good (20–25), fair (15–19), and poor (14). The checklist can be used to assess both randomised and non-randomised experiments (Downs and Black 1998).

3.2.17 GRACE Checklists

The GRACE Checklist (Good Research for Comparative Effectiveness) is an 11-item instrument for assessing the quality of data and procedures used in study design and analysis. Six criteria assess the data's quality, while five others focus on the study's design and analysis procedures (Dreyer et al. 2016).

3.3 Methodological Index for Non-randomised Studies (MINORS)

This score for evaluating non-randomised studies was created by a group of surgeons in response to physicians' concerns about the dearth of randomised surgical trials and the enormous quantity of observational studies in surgery. This is a significant factor for clinical research's 'consumers.'

MINORS, in my perspective, have two key characteristics. First, because it is simple, with only 12 components, it is easily utilised by both readers and researchers, and second, because it is reliable, as evidenced by clinimetric testing (Slim et al. 2003).

3.4 Conclusion

In conclusion, various tools are available for the assessment of the quality of studies. Therefore, researchers could select the suitable available tool depending upon the aim and objectives of the SLR.

References

Agency for Healthcare Research and Quality (2022). https://effectivehealthcare.ahrq.gov. Assessed 12 Feb 2022

Atkins D, Best D, Briss PA et al (2004) GRADE Working Group. Grading quality of evidence and strength of recommendations. BMJ 328(7454):1490

Centre for Evidence Based Medicine (CEBM) tools (2022). https://www.cebm.ox.ac.uk/resources/ebm-tools. Accessed 12 Feb 2022

Chung JH, Lee JW, Jo JK et al (2013) A quality analysis of randomized controlled trials about erectile dysfunction. World J Mens Health 31(2):157–162

Critical Appraisal Skills Programme (CASP) Checklists (2022). https://casp-uk.net/casp-tools-checklists/. Accessed 12 Feb 2022

Database of Abstracts of Reviews of Effects (DARE) (1995) Quality-assessed reviews. Centre for Reviews and Dissemination (UK), York (UK). https://www.ncbi.nlm.nih.gov/books/NBK285222/. Accessed 4 Nov 2021

Downs SH, Black N (1998) The feasibility of creating a checklist for the assessment of the methodological quality both of randomised and non-randomised studies of health care interventions. J Epidemiol Commun Health 52:377–384

Dreyer NA, Bryant A, Velentgas P (2016) The GRACE checklist: a validated assessment tool for high quality observational studies of comparative effectiveness. J Manag Care Spec Pharm 22(10):1107–1113

Jadad AR, Moore RA, Carroll D et al (1996) Assessing the quality of reports of randomized clinical trials: is blinding necessary? Control Clin Trials 17(1):1–12

Joanna Briggs Institute (JBI) (2022) Critical appraisal tools. https://jbi.global/critical-appraisal-tools Accessed 12 Feb 2022

Jørgensen L, Paludan-Müller AS, Laursen DR et al (2016) Evaluation of the Cochrane tool for assessing risk of bias in randomized clinical trials: overview of published comments and analysis of user practice in Cochrane and non-Cochrane reviews. Syst Rev 80:5

Margulis AV, Pladevall M, Riera-Guardia N et al (2014) Quality assessment of observational studies in a drug-safety systematic review, comparison of two tools: the Newcastle-Ottawa scale and the RTI item bank. Clin Epidemiol 6:359–368

Review Manager (RevMan) (2020) [Computer program]. Version 5.4. The Cochrane Collaboration

Shea BJ, Reeves BC, George W et al (2017) AMSTAR 2: a critical appraisal tool for systematic reviews that include randomised or non-randomised studies of healthcare interventions, or both. BMJ 358:j4008

Skrivankova VW, Richmond RC, Woolf B et al (2021) Strengthening the reporting of observational studies in epidemiology using Mendelian randomisation (STROBE-MR): explanation and elaboration. BMJ 375:n2233

Slim K, Nini E, Forestier D et al (2003) Methodological index for non-randomized studies (MINORS): development and validation of a new instrument. ANZ J Surg 73(9):712–716

Strengthening the reporting of observational studies in epidemiology (2022) STROBE checklists. https://www.strobe-statement.org/checklists/ Accessed 12 Feb 2022

van Tulder M, Furlan A, Bombardier C et al (2003) Updated method guidelines for systematic reviews in the Cochrane Collaboration Back Review Group. Spine (Phila Pa 1976) 28:1290–1299

von Elm E, Altman DG, Egger M et al (2007) STROBE Initiative. The strengthening the reporting of observational studies in epidemiology (STROBE) statement: guidelines for reporting observational studies. Bull World Health Organ 85(11):867–872

Whiting PF, Rutjes AW, Westwood ME et al (2011) QUADAS-2 Group. QUADAS-2: a revised tool for the quality assessment of diagnostic accuracy studies. Ann Intern Med 155(8):529–536

Whiting P, Savović J, Higgins JP et al (2016) ROBIS: a new tool to assess risk of bias in systematic reviews was developed. J Clin Epidemiol 69:225–234

Extraction and Analysis of Data

4

Abstract

The extraction of data from selected studies followed by proper analysis is one of the important steps in meta-analysis. The data should be extracted by at least two independent authors to reduce the errors. The extracted data should be analysed carefully to calculate an overall estimate with a 95% confidence interval, along with an analysis of heterogeneity and publication bias. This chapter provides details about the extraction of data from selected studies followed by analysis using suitable software.

Keywords

Meta-analysis · Overall estimate measure · Heterogeneity · Publication bias

4.1 Introduction

The data from studies is extracted carefully after completion of quality assessment, which is usually done by well-designed scales as discussed in detail in Chap. 3. Individual studies' data is retrieved in a consistent format to enable comparison between studies. The overall estimate is calculated in terms of relative risk (RR), odd ratios (ORs), standard mean difference (SMD), and mean difference (MD) along with a 95% confidence interval depending upon the type of data. The weightage is provided for individual studies (Mikolajewicz and Komarova 2019). This chapter provides detailed information regarding the extraction and analysis of data.

4.2 Extraction of Data

The data from studies should be extracted by at least two authors to minimise errors.

A. Kumar, *Meta-analysis in Clinical Research: Principles and Procedures*,
https://doi.org/10.1007/978-981-99-2370-0_4

4.3 Analysis of Data

The analysis of data should be done in a systematic way. The following points need to be considered for the analysis of data.

4.3.1 Weightage to Studies

The weightage is given to the individual studies depending upon the size of the trial. The results from studies with a small sample size are provided less weightage as compared to studies with a large sample size.

4.3.2 Selection of Model

The selection of a model is also important, which depends upon the variability between the studies. The fixed effect model is preferred if the variations between studies are small; otherwise, the random effect model is preferred. Only if studies are highly heterogeneous there will be a significant difference in the total effect calculated by the fixed and random effects models. It may not be suitable to aggregate the results if the research's outcomes are vastly different. However, it is unclear how to determine whether it is appropriate. One method is to look at the degree of similarity in the studies' outcomes statistically. In other words, to look for heterogeneity between studies, the results of a study are evaluated to see if they indicate a single underlying effect rather than a dispersion of effects. If the findings of the test are homogeneous, the discrepancies between studies are presumed due to sampling variance, and a fixed effect model is suitable. Conversely, the test reveals high heterogeneity between study outcomes, a random effects model is recommended. However, many researchers are not in favour of selecting a model based upon the heterogeneity among studies.

4.3.3 Choose an Effect Size

The selection of an effect size index should be guided by three primary concerns. The first is that the effect sizes from multiple studies should be comparable, meaning that they should measure the same thing. That is, the amount of the effect should not be influenced by characteristics of research design that may differ from one study to the next (such as sample size or whether covariates are used). The second point is that impact size estimates should be calculated using information that is likely to be disclosed in published study papers. That is, the raw data should not have to be re-analysed (unless these are known to be available). The final requirement is that the effect size be technically sound. Its sample distribution, for example, needs to be understood in order to determine variances and confidence intervals. Furthermore, the effect size should be meaningfully interpretable. This means that the effect size should be useful to researchers in the substantive field of the study covered in the synthesis. It is usually easy to change the effect size to another metric for presentation if it is not inherently significant. The analysis might, for example, be carried out using the log risk ratio, but then translated into a risk ratio (or even illustrated risks) for presentation. The effect size should also have appropriate technical features, according to the third point. Its sample distribution must be understood in order to compute variances and confidence intervals. Furthermore, the effect size must be interpretable in a meaningful way. This implies that the effect size should be meaningful to researchers working in the substantive field of the study as reflected in the synthesis.

In practice, the type of data utilised in primary research frequently results in a pool of two or three effect sizes that match the criteria mentioned above, making the process of choosing an effect size quite simple. The raw difference in mean, or the standardised difference in mean, will usually be the appropriate effect size if the summary statistics presented by the primary study are based on mean and standard deviations in two groups. If the summary data is based on a binary result, such as events and non-events in two groups, the risk ratio, odds ratio, or risk difference will usually be the appropriate impact size. If a correlation between two variables is reported in the primary study, the correlation coefficient may be used as the effect size. The overall estimate is also calculated as prevalence, for example, might be performed to combine multiple estimates of the prevalence of neurological symptoms in COVID-19 patients. The overall estimate is calculated as a raw difference in mean when the outcome is presented on a meaningful scale and all studies in the analysis utilise the same scale. The raw mean difference has the benefit of being immediately intelligible, either because of its inherent meaning (for example, blood pressure is measured on a known scale) or because of its widespread use (for example, a national achievement test for students, where all relevant parties are familiar with the scale). Consider a study that offers mean values for two groups (treated and control), and we want to compare the two groups' mean. The population mean difference is defined as the difference between the mean of the treated and control groups. We can calculate the standard deviation.

When a researcher has access to a complete set of summary data for each group, such as the mean, standard deviation, and sample size, computing the effect size and variance is quite simple. However, in actuality, the researchers are frequently working with only a portion of the data. For example, a study may just publish the p-value, mean, and sample sizes from a test of significance, leaving the effect size and variance computation to the meta-analyst.

4.3.4 Mean Difference (MD) vs Standardised Mean Difference (SMD)

The raw mean differences are usually combined if the same scale is used for the measurement of the outcome across the studies. However, it is not appropriate to combine if different studies use different scales for the assessment of the outcome. In such circumstances, we may generate a similar index (the standardised mean difference) by dividing the mean difference in each study by the standard deviation of that study. The standardised mean difference converts all effect sizes into a single metric, allowing us to include many outcome measures in a single synthesis.

Overall, the mean difference is calculated if included studies have used the same scale for the measurement of the outcome, whereas a standardised mean difference is usually calculated if studies have used different scales for the measurement (Andrade 2020; Takeshima et al. 2014; Sedgwick and Marston 2013).

4.3.5 Response Ratios

The ratio of the mean in the two groups could be used as an impact size index in research domains where the outcome is assessed on a physical scale (such as length, area, or mass) and is unlikely to be zero. The response ratio is the name given to this effect size metric in experimental ecology. The computations for response ratios are done on a log scale. We calculate the log response ratio and its standard error, then utilise these data to complete all steps of the meta-analysis, and ultimately convert the results back to the original metric (Bakbergenuly et al. 2020).

4.3.5.1 Effect Sizes Based on Binary Data

As an overall estimation measure for binary data, the relative risk, odd ratio, and risk differences are determined. We can calculate the probability of an occurrence (such as death) in each group (for example, treated versus control). The ratio of these hazards is then used to determine the effect size (the risk ratio). We can calculate the probability of an event (such as the death-to-survival ratio) in each group (for example, treated versus control). The odds ratio is then calculated as the ratio of these odds. We can calculate the probability of an occurrence (such as death) in each group (for example, treated versus control). The difference between these risks is then used to calculate the effect size (the risk difference). As the first step to work with the risk ratio or odds ratio, researchers should transform all values into log values. Next is to perform the analyses and convert the results back to ratio values. To work with the risk difference, we work with the raw values. These effect sizes are explored in detail below, along with appropriate examples.

4.3.5.1.1 Risk Ratio

Simply said, the risk ratio is the ratio of two risks. Assume the treated group's risk of death is 5/100 and the control group's risk of death is 10/100, resulting in a risk ratio of 0.50. The advantage of this index is that it is intuitive, in the sense that the meaning of a ratio is obvious. Calculations for risk ratios are done on a log scale. We compute the log risk ratio, and the standard error of the log risk ratio, and use these numbers to perform all steps in the meta-analysis. Only then we can convert the results back into the original metric. The log transformation is needed to maintain symmetry in the analysis.

4.3.5.1.2 Odds Ratio

The risk ratio is the ratio of two risks, whereas the odds ratio is the ratio of two odds. Let us continue this with the example mentioned in the above section of the risk ratio, the odds of death in the treated group would be 5/95, or 0.0526 (since the probability of death in the treated group is 5/100 and the probability of life is 95/100), while the odds of death in the control group would be 10/90, or 0.1111 (since the probability of death in the treated group is 5/100 and the probability of life is 95/100). 0.0526/0.1111, or 0.4737, would be the ratio of the two odds. Although the odds ratio has statistical qualities that make it the best choice for a meta-analysis, many individuals find it less obvious than the risk ratio. The computations for odds ratios are done on a log scale (for the same reason as for risk ratios). We calculate the log odds ratio and the standard error of the log odds ratio, and we will need these data to complete all of the meta-analysis procedures. Finally, we have to convert the findings back to the original measure.

4.3.5.1.3 Risk Difference

The difference between two risks is referred to as the risk difference. In continuation of the above example, the risk in the treatment group is 0.05, whereas the risk in the control group is 0.10, resulting in a 0.05 risk difference. The risk differences are calculated in raw units rather than log units, whereas the risk ratios and odds ratios are calculated in log units (Viera 2008; Ranganathan et al. 2015).

4.4 Selection of Effect Sizes (Risk Ratio, Odds Ratio, and Risk Difference)

The researcher must evaluate both substantive and technical aspects when choosing between the risk ratio, odds ratio, and risk difference. Because the risk ratio and the odds ratio are relative measures, they are insensitive to variations in baseline occurrences. The risk difference, however, is an absolute measure that is highly sensitive to the baseline risk. If we want to test a chemical and felt it decreased the risk of an event by 20% regardless of the baseline risk, we would expect to see the same effect size across trials using a ratio index, even if the baseline risk differed from the study to study. In studies with a higher base rate, however, the risk difference would be greater.

Conversely, if we wanted to express the treatment's clinical benefit, the risk difference might be a better indicator. Assume we conduct a meta-analysis to compare the risk of adverse events in the treatment and control groups. The risk for treated patients is 1/1000 compared to 1/2000 for control patients, resulting in a risk ratio of 2.00. Simultaneously, the risk difference is 0.0010 versus 0.0005 for a 0.0005 risk difference. Both of these values (2.00 and 0.0005) are valid, yet they represent two different things. Some recommend utilising the risk ratio (or odds ratio) to perform the meta-analysis and compute a summary risk (or odds) ratio since the ratios are less susceptible to baseline risk while the risk difference is sometimes more clinically significant.

4.4.1 Effect Sizes Based on Correlations

When research offers data as correlations, we commonly use the correlation coefficient itself as the effect magnitude. We use the Fisher's z transformation to change the correlation and then use this index to analyse it. The summary results are then converted back to correlations for presentation.

4.4.2 Converting Among Effect Sizes

If all of the studies in the analysis use the same type of data (means, binary, or correlational), the researcher should use that data to choose an impact size. We can use formulas to translate between effect sizes when some studies utilise means, others use binary data, and yet others use correlational data. Research that utilises different measures may differ in significant ways, and we must evaluate this possibility when considering whether it is appropriate to include the studies in the same analysis.

If some studies indicate a difference in means, which is used to calculate a standardised mean difference, while others report a difference in proportions, which is used to calculate an odds ratio, and yet others show a correlation, the standardised mean difference will be computed. We want to include all of the studies into one meta-analysis because they all address the same broad subject. As a result, they must be converted to a common effect size. When effect size conversion is not achievable, the question of whether or not it is permissible to mix effect sizes from studies using various metrics must be examined on a case-by-case basis.

Assume that numerous randomised controlled trials begin with the same continuous scale measure, but that some publish the result as a mean, while others dichotomize the result and present it as success or failure. In this scenario, converting the standardised mean differences and odds ratios to a common metric and then combining them across studies may be a good idea. Observational studies that show correlations, however, may differ significantly from observational studies that report odds ratios. Even though there is no technological barrier to converting the effects to a common metre in this scenario, it might not be a good idea from a substantive standpoint.

4.4.3 Calculation of Heterogeneity

The heterogeneity among studies should be calculated and examined further to understand why treatment effects differ in different circumstances? (Higgins et al. 2003).

4.5 Conclusion

In conclusion, the extraction of data from good quality studies should be done in a well-designed format as per the aim and objective of the review. The overall estimate can be calculated further depending upon the type of data. Various web-based tools are also available which help in the screening and extraction of data. DistillerSR is a web-based software tool for screening and extracting data from bibliographic records. It contains a number of management capabilities that allow you to keep track of your work, evaluate interrater reliability, and export data for future research (DistillerSR Smarter Reviews: Trusted Evidence 2022). EPPI-Reviewer is a web-based application that helps with reference management, screening, risk of bias assessment, data extraction, and synthesis during the systematic review process. This tool is also available for purchase. Only a free trial is provided. Rayyan is a collaborative citation screening and full-text selection web tool. This tool is free to use (EPPI-Reviewer 2022).

References

Andrade C (2020) Mean difference, standardized mean difference (SMD), and their use in meta-analysis: as simple as it gets. J Clin Psychiatry 81(5):20f13681

Bakbergenuly I, Hoaglin DC, Kulinskaya E (2020) Estimation in meta-analyses of response ratios. BMC Med Res Methodol 20(1):1–24

DistillerSR Smarter Reviews: Trusted Evidence (2022). https://www.evidencepartners.com/products/distillersr-systematic-review-software. Accessed 10 Jan 2022

EPPI-Reviewer (2022). https://eppi.ioe.ac.uk. Accessed 10 Jan 2022

Higgins JP, Thompson S, Deeks JJ et al (2003) Measuring inconsistency in meta-analyses BMJ (Clinical research ed) 327(7414):557–560

Mikolajewicz N, Komarova SV (2019) Meta-analytic methodology for basic research: a practical guide. Front Physiol 10:203

Ranganathan P, Aggarwal R, Pramesh CS (2015) Common pitfalls in statistical analysis: odds versus risk. Perspect Clin Res 6(4):222–224

Sedgwick P, Marston L (2013) Meta-analyses: standardised mean differences. BMJ 347:f7257

Takeshima N, Sozu T, Tajika A et al (2014) Which is more generalizable, powerful and interpretable in meta-analyses, mean difference or standardized mean difference? BMC Med Res Methodol 14(1):1–7

Viera AJ (2008) Odds ratios and risk ratios: what's the difference and why does it matter? South Med J 101(7):730–734

Models

5

Abstract

The most commonly used models in the meta-analysis are fixed and random effect models. The choice of a specific model is critical to properly analysing the data. This chapter started with the introduction of fixed and random effect models. The selection of a particular model is discussed in detail.

Keywords

Fixed effect models · Random effect models

5.1 Introduction

The fixed and random effect models are the most commonly used models in the meta-analysis. The main differences between the assumptions of the fixed and random effect models are differences in true effects. In the fixed effect model, we assume the same effect size in all included studies, whereas, in the random effect model, we assume variations in effect sizes across the studies. The main goal of the fixed effect model is to calculate one true effect, whereas the goal of the random effect model is to calculate the mean of the distribution of effects. Therefore, we cannot exclude a small study by giving it less weightage and too much weightage to a very large study (Borenstein et al. 2010). In this chapter, we have discussed in detail the fixed and random effect model in meta-analysis.

A. Kumar, *Meta-analysis in Clinical Research: Principles and Procedures*,
https://doi.org/10.1007/978-981-99-2370-0_5

5.2 How Will the Selection of a Model Influence the Overall Effect Size?

Let us understand it with a suitable example. Assume the author has conducted a meta-analysis of individual studies data to find out the role of aspirin in the reduction of mortality of COVID-19 patients and the results are presented in the forest plot (Fig. 5.1). In this particular example, the highest effect size and high weightage (16.2%) are provided in the large study, i.e. Lodigiani et al. (2020), in the fixed effect model as shown in Fig. 5.1. By contrast, under the random effect model, Lodigiani et al. (2020) study was given a modest weightage (13.8%). Therefore, there is a less effect on the pooled estimate as shown in Fig. 5.2.

Similarly, Viecca et al. (2020) smallest study had the smallest effect size. Under the fixed effect model, Viecca et al. (2020) assigned less weightage and had less influence on effect size. By contrast, in the random effect model, Viecca et al. (2020) got a somewhat higher weightage. As a result, relative weights awarded using the random effect model are more evenly distributed than those allocated using the fixed

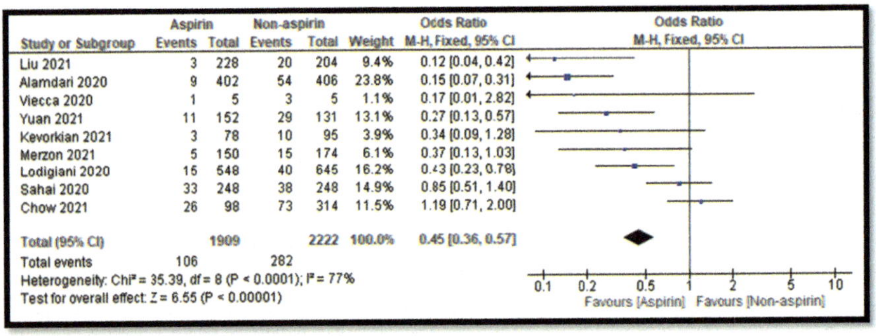

Study or Subgroup	Aspirin Events	Aspirin Total	Non-aspirin Events	Non-aspirin Total	Weight	Odds Ratio M-H, Fixed, 95% CI
Liu 2021	3	228	20	204	9.4%	0.12 [0.04, 0.42]
Alamdari 2020	9	402	54	406	23.8%	0.15 [0.07, 0.31]
Viecca 2020	1	5	3	5	1.1%	0.17 [0.01, 2.82]
Yuan 2021	11	152	29	131	13.1%	0.27 [0.13, 0.57]
Kevorkian 2021	3	78	10	95	3.9%	0.34 [0.09, 1.28]
Merzon 2021	5	150	15	174	6.1%	0.37 [0.13, 1.03]
Lodigiani 2020	15	548	40	645	16.2%	0.43 [0.23, 0.78]
Sahai 2020	33	248	38	248	14.9%	0.85 [0.51, 1.40]
Chow 2021	26	98	73	314	11.5%	1.19 [0.71, 2.00]
Total (95% CI)		1909		2222	100.0%	0.45 [0.36, 0.57]
Total events	106		282			

Heterogeneity: $Chi^2 = 35.39$, $df = 8$ ($P < 0.0001$); $I^2 = 77\%$
Test for overall effect: $Z = 6.55$ ($P < 0.00001$)

Fig. 5.1 Forest plot presenting the role of aspirin in COVID-19 patients using a fixed effect model

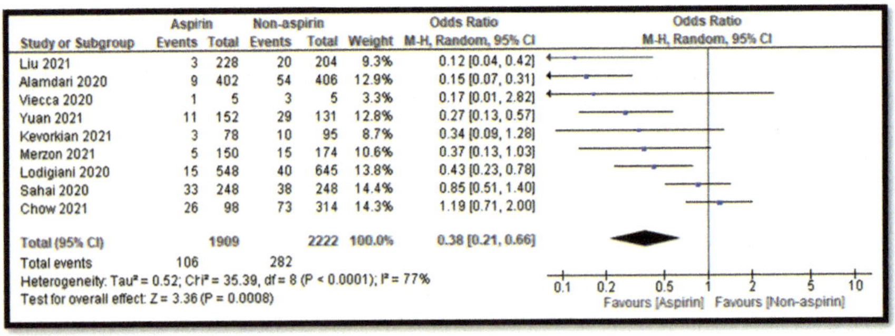

Study or Subgroup	Aspirin Events	Aspirin Total	Non-aspirin Events	Non-aspirin Total	Weight	Odds Ratio M-H, Random, 95% CI
Liu 2021	3	228	20	204	9.3%	0.12 [0.04, 0.42]
Alamdari 2020	9	402	54	406	12.9%	0.15 [0.07, 0.31]
Viecca 2020	1	5	3	5	3.3%	0.17 [0.01, 2.82]
Yuan 2021	11	152	29	131	12.8%	0.27 [0.13, 0.57]
Kevorkian 2021	3	78	10	95	8.7%	0.34 [0.09, 1.28]
Merzon 2021	5	150	15	174	10.6%	0.37 [0.13, 1.03]
Lodigiani 2020	15	548	40	645	13.8%	0.43 [0.23, 0.78]
Sahai 2020	33	248	38	248	14.4%	0.85 [0.51, 1.40]
Chow 2021	26	98	73	314	14.3%	1.19 [0.71, 2.00]
Total (95% CI)		1909		2222	100.0%	0.38 [0.21, 0.66]
Total events	106		282			

Heterogeneity: $Tau^2 = 0.52$; $Chi^2 = 35.39$, $df = 8$ ($P < 0.0001$); $I^2 = 77\%$
Test for overall effect: $Z = 3.36$ ($P = 0.0008$)

Fig. 5.2 Forest plot presenting the role of aspirin in COVID-19 patients using a random effect model

effect model. Extreme studies will lose influence if they are huge and gain influence if they are tiny when we switch from a fixed effect to a random effect model.

Figures 5.1 and 5.2 show the size of the box reflects the weightage assigned to each study. If you look carefully at both the figures, there is a wide range of weights (size of boxes), whereas, under the random effect model, the weights fall into a relatively narrow range. Overall, study weights are more balanced under random effect models (3.3–14.4%) than under fixed effect models (1.1–23.8%). Fewer relative weights are assigned for large studies, whereas more relative weights are assigned for small studies.

5.3 Fixed Effect Model

In the fixed effect model, we assume all included studies share a common effect size, and all the factors that could influence effect sizes are same in all studies. However, in many studies, this assumption is implausible. Of course, when we plan to perform a meta-analysis, we assume the included studies have common characteristics that make sense to synthesise the information. However, we cannot assume that true effect sizes across the studies will also be identical (Kelley and Kelley 2012; Borenstein et al. 2010). For example, suppose that we are working with studies that compare the reduction of deaths of COVID-19 patients in the steroidal group and the non-steroidal group. If the treatment works, we would expect the effect size to be similar but not identical across studies. The effect size could be affected by various factors such as age, condition, sex, and so on. We might not have assessed these covariates in each study and may be even do not know which covariates are actually related to the size of the effect. Therefore, in these cases, the fixed effect model might not be suitable.

5.4 Random Effect Model

In both models (fixed and random effect), each study was weighted by the inverse of its variance. A random effect model is preferred if the variance among studies is high. There are generally two sources of variance (within and between studies) that should be considered.

5.5 Confidence Interval

The only source of variation under the fixed effect model is within the study, whereas under the random effect model, the source of variations is within studies as well as between the studies. Therefore, variance, standard error, and confidence interval for summary effect are always wider in the random effect model as compared to the fixed effect model. In this example, the confidence interval for the summary effect in the random effect model is 0.38 [0.21, 0.66], whereas under the fixed effect model it is 0.45 [0.36, 0.57] as shown in Figs. 5.1 and 5.2.

5.6 Which Model Should We Use?

A number of researchers are perplexed when it comes to selecting a model. Many researchers are starting with a fixed effect model and then move to a random effect model if statistical tests indicate high heterogeneity among studies. In my personal opinion, it is a wrong practice and should be discouraged. The decision to use which model should be based on our understanding about the common effect sizes, not on the outcome of a statistical test for heterogeneity, as it often suffers from low power.

Therefore, the selection of a model should be done at the design phase of the meta-analysis. The fixed effect model should be preferred if all included studies are functionally identical and the computed effect size does not generalise to other populations. For example, a Phase 3 trial of a drug will use thousands of patients to compare the effects of the interventional drug with a standard drug. However, if the trial is divided into various stages, only 100 patients at a time, and the sponsor will run 10 trials for each of those 100 patients. The studies are assumed to be similar as the effects of variables on outcome are the same across 10 studies. The same researchers have conducted all these 10 studies with the same dose, duration, and so on and all studies are expected to share a common effect. Thus, the first condition is met to use a fixed effect model. If the results of these studies cannot be extrapolated to other populations. So, in this particular case, a fixed effect model should be used instead of a random effect model.

In most cases, it would be unlikely that the included studies were functionally equivalent. These studies vary among each other, which would have an impact on the results such as age, sample size, design of the study, and so on. Therefore, in such cases, the random effect model should be preferred (Schmidt et al. 2009; Nikolakopoulou et al. 2014).

5.7 Conclusion

The fixed and random effect models are the most commonly used in meta-analysis and should be chosen based on the variations within and between studies rather than the results of the statistical tests.

References

Borenstein M, Hedges LV, Higgins JP et al (2010) A basic introduction to fixed-effect and random-effects models for meta-analysis. Res Synth Methods 1(2):97–111

Kelley GA, Kelley KS (2012) Statistical models for meta-analysis: a brief tutorial. World J Methodol 2(4):27

Lodigiani C, Iapichino G, Carenzo L, et al (2020) Venous and arterial thromboembolic complications in COVID-19 patients admitted to an academic hospital in Milan, Italy. Thromb Res 191:9–14

Nikolakopoulou A, Mavridis D, Salanti G (2014) How to interpret meta-analysis models: fixed effect and random effects meta-analyses. Evid Based Ment Health 17(2):64

Schmidt FL, Oh IS, Hayes TL (2009) Fixed-versus random-effects models in meta-analysis: model properties and an empirical comparison of differences in results. Br J Math Stat Psychol 62(1): 97–128

Viecca M, Radovanovic D, Forleo GB, Santus P (2020) Enhanced platelet inhibition treatment improves hypoxemia in patients with severe Covid-19 and hypercoagulability. A case control, proof of concept study. Pharmacol Res 158:104950

Heterogeneity and Publication Bias

6

Abstract

In a meta-analysis, heterogeneity refers to the differences in research outcomes between studies. Heterogeneity is not something to be terrified of; it simply implies that your data is variable. When multiple research projects are brought together for analysis or a meta-analysis, it is apparent that differences will be discovered. The only thing is to identify and analyse the heterogeneity in a proper way. Publication bias is also very important to be assessed qualitatively and quantitatively. This chapter provides a detailed discussion regarding heterogeneity and publication bias.

Keywords

Heterogeneity · Publication bias · Meta-analysis

6.1 Introduction

A meta-analysis attempts to provide unbiased answers. The classic systematic review addresses a question related to the effectiveness of treatment in different groups and usually includes only randomised controlled trials (RCTs). However, if very few RCTs are available, other types of studies might also be included, which could result in selection bias. Therefore, included studies should be systematically appraised for the risk of bias. The goal of a meta-analysis is not simply to calculate the overall estimate but, most importantly, to make sense of the pattern of effects. We should consider various important factors that directly impact the results. Heterogeneity refers to variations among studies. It is not something to be afraid of, it just means there is variability among studies that are included in the analysis. It is obvious that when we pool data from individual studies, heterogeneity will exist across the studies. The heterogeneity among studies is broadly categorised into three

A. Kumar, *Meta-analysis in Clinical Research: Principles and Procedures*, https://doi.org/10.1007/978-981-99-2370-0_6

categories: clinical (differences in participants, interventions, or outcomes), methodological (differences in study design, risk of bias), and statistical (variation in intervention effects or results). Some researchers criticise meta-analysis as it looks like mixings apples and oranges. It combines the results of individual studies, which are broadly similar but can differ in a number of ways, such as age, gender, dosage, treatment schedule, etc. However, the assumption of the meta-analysis approach is that if a treatment is effective, then the effect of treatment in different trials will go in the same direction, and heterogeneity among studies should be measured and subgroup analysis should be conducted to check the effect of individual factors on the results. Heterogeneity tests are widely used to decide on strategies for merging studies and determine whether or not the findings are consistent. Meta-analysis requires a thorough examination of the consistency of effects across studies. We cannot establish the generalisability of meta-analysis conclusions unless we know how consistent the outcomes of investigations are. Indeed, several hierarchical systems for grading evidence say that for the highest grade, study results must be consistent or homogeneous (Fletcher 2007).

In systematic reviews and meta-analyses, publication bias is a severe problem that can impair the validity and generalisation of conclusions. A meta-analysis, which summarises quantitative evidence from various studies, is increasingly employed in a variety of scientific areas. If the synthesised literature has been tainted by publication bias, the meta-analytic results will be tainted as well, perhaps leading to inflated conclusions. Publication bias in meta-analysis covers the various types of publication bias and proposes strategies for evaluating and lowering, or even eliminating, publication bias.

The failure to publish the results of specific studies due to the direction, nature, or strength of the study findings is known as publication bias. Outcome-reporting bias, time-lag bias, grey-literature bias, full-publication bias, language bias, citation bias, and media-attention bias are all examples of publishing bias in academic articles. According to reports, more than 20% of finished studies may not be published due to a variety of factors, including publication bias. Studies with limited sample size, as well as those with non-significant or negative findings, are less likely to be published, particularly in high-impact journals. Studies with non-significant results, however, are significantly more delayed in being published than those with significant results. Furthermore, research undertaken outside of English-speaking countries is less likely to be published in English-language peer-reviewed journals. As a result, results from published studies may be consistently different from those unpublished investigations, posing problems for meta-analysis (Deeks et al. 2009). This chapter discussed in detail regarding heterogeneity and publication bias in meta-analysis.

6.2 How to Identify and Measure Heterogeneity?

The identification of heterogeneity is an important step followed by its measurement. Some tests are available for this purpose which are explained below with suitable examples.

6.2.1 Eyeball Test

The eyeball test is a simpler test to identify heterogeneity among studies. To understand in a much simpler way, I have included two forest plots as presented in Figs. 6.1 and 6.2. Just look carefully at the overlapping confidence intervals of individual studies rather than which side effect estimates are. The less overlapping confidence intervals of different studies indicate more heterogeneity. Therefore, Fig. 6.2 shows more heterogeneity among studies as compared to Fig. 6.1, as less confidence intervals are overlapping and in addition to that, all studies favour control

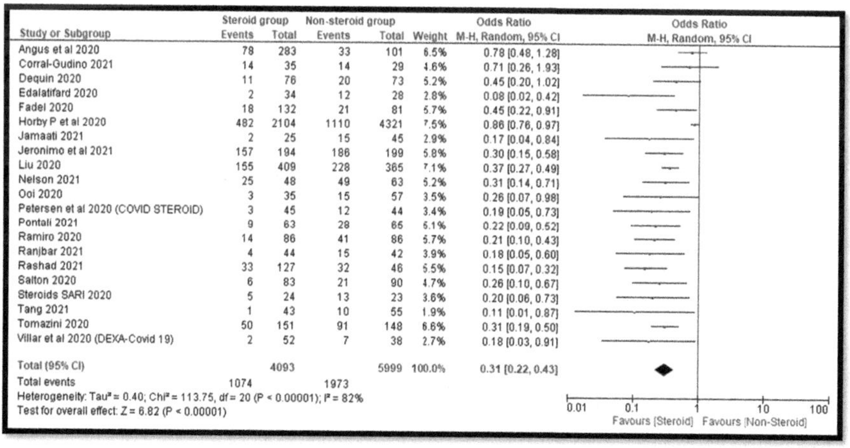

Fig. 6.1 Forest plot showing the role of steroids in COVID-19 patients

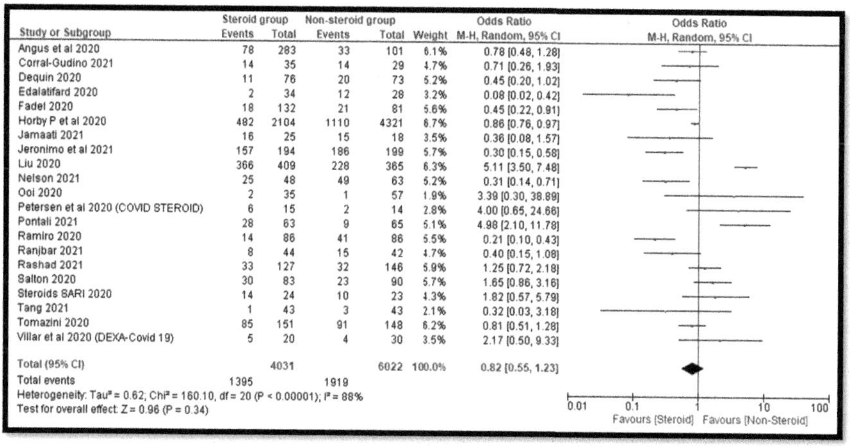

Fig. 6.2 Forest plot showing the role of steroids in COVID-19 patients

intervention. Apart from the eyeball test, we have some specific statistical methods, as described in further sections which help to measure heterogeneity among studies.

6.2.2 Chi-Squared (χ^2) Test

The Chi-squared (χ^2) test provides a *p*-value to test the hypothesis. The low *p*-value indicates rejection of the null hypothesis, i.e. each study measuring similar effects and heterogeneity among studies is low. Therefore, a low *p*-value indicates heterogeneity among studies. The cut-off *p*-value is <0.05. However, many researchers consider <0.1 instead of <0.05 as the cut-off *p*-value due to the low sensitivity of the Chi-squared (χ^2) test (Lin et al. 2017).

6.2.3 I^2

This test measures the extent of heterogeneity rather than only stating if it is present or not, as shown by the eyeball test and the Chi-squared (χ^2) test. There is no specific threshold as the heterogeneity depends upon various factors (von Hippel 2015). A rough range regarding heterogeneity is mentioned below.

0–40%: low heterogeneity
30–60%: moderate heterogeneity
50–90%: substantial heterogeneity
75–100%: considerable heterogeneity

To understand the I^2 and Chi-squared (χ^2) test, I have included a suitable example as shown in Fig. 6.3. In this example, the *p*-value of the Chi-squared (χ^2) test is 0.11 which confirms the acceptance of the null hypothesis (homogeneity). However, I^2 indicates 51% (moderate) heterogeneity. I^2 has a substantial bias when the number of studies is small.

Therefore, Chi-squared (χ^2) test can be misleading, particularly when few studies are included for analysis.

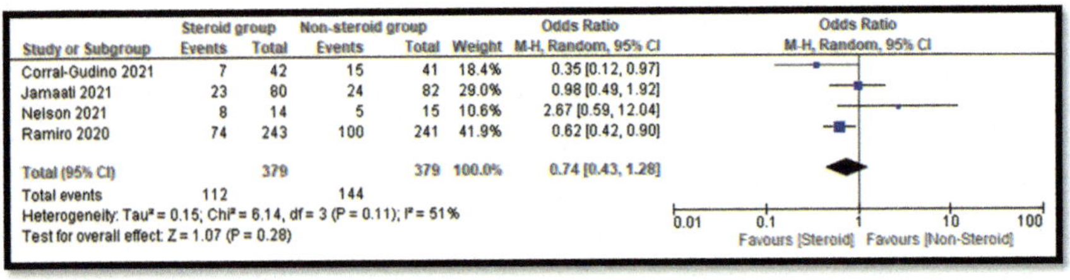

Fig. 6.3 Forest plot showing heterogeneity results

6.2.4 Cochran's *Q* Test

The heterogeneity is also determined using Cochran's Q test, which evaluates the weighted sum of squared differences between individual study effects and the pooled effect across trials. With k (number of studies) minus 1 degree of freedom, Q is distributed as a chi-square statistic. Q has low power in the case of a smaller number of studies and vice versa (Cohen et al. 2015; Kulinskaya and Dollinger 2015).

6.3 How to Deal with Heterogeneity?

If you have observed high heterogeneity in your analysis, you need to deal with it. Some of the steps are mentioned below, which help researchers to deal with heterogeneity.

1. The data should be cross-checked for mistakes. Mostly, high heterogeneity is observed due to typographical errors.
2. If heterogeneity is still too high (after careful cross-checking of data), my personal advice to researchers do not go for meta-analysis, you can restrict yourself to SLR.
3. If heterogeneity is too high, researchers should perform a meta-regression to explore heterogeneity.

6.4 Publication Bias

The publication bias is important to analyse. Currently, we are doing qualitative and quantitative analyses of publication bias. Visual inspection of funnel plots and non-parametric trim and fill methods are used in qualitative analysis, whereas weight functions are used in quantitative analysis to adjust the overall effect size estimate and are commonly used as sensitivity analyses to assess the potential impact of publication bias (Philip 2015; Duval and Tweedie 2000).

6.4.1 Assessment of Publication Bias

The publication bias is assessed qualitatively using a funnel plot and quantitatively using the Begg and Egger test, as described below.

6.4.1.1 The Funnel Plot

A funnel plot is one of the most extensively used methods for determining publication bias. A funnel plot is a scatterplot of estimated treatment effects from individual studies versus study precision measurements. Because a study's accuracy of effect estimate is proportional to its sample size, larger studies with higher precision will appear at the top of a funnel plot, while smaller studies will appear at the bottom.

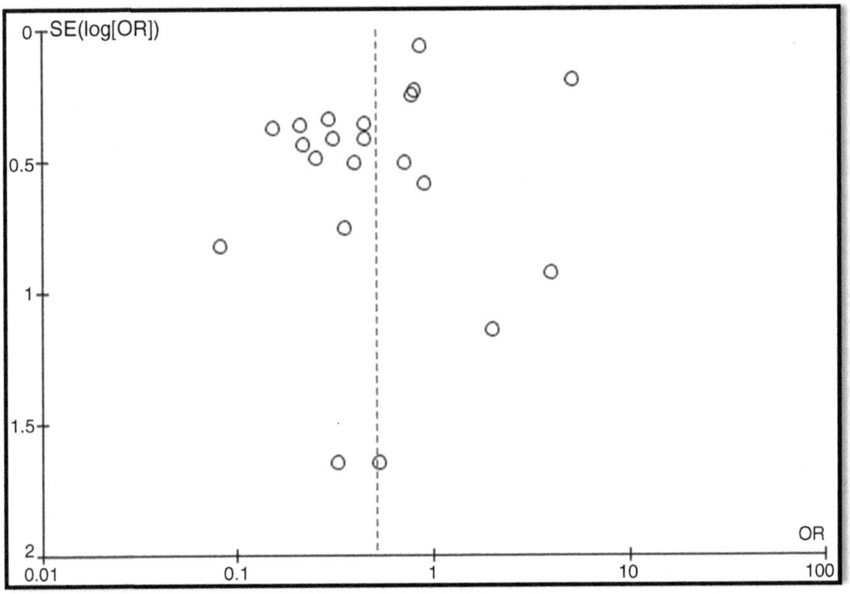

Fig. 6.4 Funnel plot

Furthermore, because smaller studies frequently represent more narrow and homogeneous sub-populations, effect estimates from smaller studies often have a wide range and are less reliable when compared to the general population. If all of the studies are representative samples from the same population, the plot should resemble a symmetrical, inverted funnel with a small top and a wider bottom (Fig. 6.4).

If there is publication bias, studies with non-significant results will be excluded from the funnel plot due to a lack of availability in the literature. Additionally, the funnel plot will have a gap on one side. The studies represented in the sample plot, for example, will not be included (Fig. 6.1). Notably, in the latter case, meta-analysis results will overestimate genuine effect sizes, with more considerable bias resulting in a more pronounced overestimation (Duval and Tweedie 2000).

It is also worth mentioning that publication bias is not the primary reason for asymmetry in funnel plots. In reality, asymmetry in a funnel plot is a better indicator of a small study effect, because studies with a small sample size are more likely to have a diverse, and typically wider, range of effect sizes than studies with a larger sample size.

Despite the fact that a funnel plot is a useful tool for assessing potential publication bias, studies have shown that many researchers are unable to visually detect publication bias using such a plot, and that the same image can be perceived differently by various researchers.

6.4.1.2 Tests for Assessing Funnel Plot Asymmetry

The two most commonly used tests for determining funnel plot asymmetry are the Begg's and Egger's tests.

6.4.1.2.1 The Begg's Rank Test

The Spearman correlation between corrected effect sizes and their variances was used to build the test presented by Begg and Mazumbar. The funnel plot asymmetry is indicated by the correlation's divergence from zero. In other words, because both the effect size and the variance of the estimate are bigger for smaller studies, a positive correlation between the effect size and the variance of the estimate is expected in the presence of publication bias.

6.4.1.2.2 The Egger's Test

Egger et al. (1997) (Duval and Tweedie 2000) advocated comparing the intercept from the regression of standard normal deviation versus precision to determine the degree of funnel plot asymmetry. The standard normal deviation (SND), which is defined as the odds ratio split by its standard error, is regressed against the accuracy of the estimate. Due to increased standard errors, both precision and SNDs are tiny in smaller trials; however, precision is large in larger studies, and if the treatment effects are significant, the SNDs are likewise large. As a result, the regression line will scatter around a line that goes through the origin for studies that represent randomly selected samples from a population, with the slope b reflecting the amount and direction of the effect. Otherwise, the regression line will be far from the origin if there is asymmetry.

In general, the Egger's test has more power than the Begg's test, the power of both tests are modest, particularly when effect sizes are variable among trials. Furthermore, because both measures are designed to differentiate between large and small studies, neither would perform well if all of the studies included were large or small (Lin and Chu 2018; Ayorinde et al. 2020).

6.5 How to Avoid Publication Bias?

Although it is impossible to totally avoid publication bias, efforts have been made to reduce the possibility of publication bias.

6.5.1 Prospective Registration

Many initiatives have been made to encourage prospective clinical study registration. To promote transparency and research integrity, as well as to prevent selective reporting and publication bias in clinical studies, the International Committee of Medical Journal Editors mandated that, beginning in July 2005, all clinical trials are registered at or before the enrolment of the first participant as a condition of consideration for publication.

6.5.2 Search for Unpublished Results

A thorough literature search is essential to avoid publication bias in general. Unpublished results can be found by looking through sources like meeting abstracts, Ph.D. dissertations, additional materials of a published article, and contacting authors and companies/organisations directly involved in the study. Furthermore, to minimise possible language bias, a literature search should not be limited to studies published in the language of publication.

It should be noted that including unpublished results should be done with caution. Small and unpublished research is more likely to have poor study design as well as a lack of analytic and scientific rigour. To avoid introducing bias due to low study quality, unpublished results should be properly evaluated before being included.

6.5.3 Improve Publication Guidelines

The novelty or importance of the research, as well as the significance of the findings, are frequently the key reasons for approving an article for publication. To reduce publication bias, some journals have begun to use scientific and technical quality as one of the acceptance criteria rather than the significance of findings, and this improvement in publication guidelines allow authors to submit more non-significant or negative results for publication.

In conclusion, in a meta-analysis, the issue of publication bias is not minor. A meta-analysis may produce misleading, often over-estimated results if there is publication bias. A visual evaluation of a funnel plot can be used to determine publication bias, followed by a formal asymmetry test. When there is asymmetry, however, publication bias may not be the only factor. The danger of publication bias can be mitigated in part by conducting a thorough literature search that includes both published and unpublished data.

Unpublished results, which are often obtained from smaller research, should be included with special attention to ensure that these studies are methodologically sound. The majority of journals prioritise innovative and noteworthy discoveries for acceptance, and there is a trend towards more publications emphasising scientific and technical excellence. To address publication bias, a comprehensive approach is required.

6.6 Conclusion

In conclusion, heterogeneity among studies must be assessed to find out the reasons for it if it is high. Publication bias should be assessed qualitatively as well as quantitatively and efforts should be made to reduce it as much as possible.

References

Ayorinde AA, Williams I, Mannion R et al (2020) Assessment of publication bias and outcome reporting bias in systematic reviews of health services and delivery research: a meta-epidemiological study. PLoS One 15(1):e0227580

Cohen JF, Chalumeau M, Cohen R et al (2015) Cochran's Q test was useful to assess heterogeneity in likelihood ratios in studies of diagnostic accuracy. J Clin Epidemiol 68(3):299–306

Deeks JJ, Higgins JPT, Altman DG (2009) Analysing data and undertaking meta-analyses. Cochrane handbook for systematic reviews of interventions version 5.1.0. The Cochrane Collaboration

Duval S, Tweedie R (2000) Trim and fill: a simple funnel-plot-based method of testing and adjusting for publication bias in meta-analysis. Biometrics 56(2):455–463

Egger M, Davey Smith G, Schneider M, Minder C (1997) Bias in meta-analysis detected by a simple, graphical test. BMJ 315:629–634

Fletcher J (2007) What is heterogeneity and is it important? BMJ 334:94

Kulinskaya E, Dollinger MB (2015) An accurate test for homogeneity of odds ratios based on Cochran's Q-statistic. BMC Med Res Methodol 15:49

Lin L, Chu H (2018) Quantifying publication bias in meta-analysis. Biometrics 74(3):785–794

Lin L, Chu H, Hodges JS (2017) Alternative measures of between-study heterogeneity in meta-analysis: reducing the impact of outlying studies. Biometrics 73(1):156–166

Philip S (2015) What is publication bias in a meta-analysis? BMJ 351:h4419

von Hippel PT (2015) The heterogeneity statistic I2 can be biased in small meta-analyses. BMC Med Res Methodol 15:35

Bias in Meta-Analysis

7

Abstract

There are multiple biases that may influence the results of a meta-analysis. Among them, the researcher usually considers the publication bias, while all the other biases are generally underestimated or ignored. This chapter provides a piece of detailed information regarding all the possible biases in the meta-analysis and how can we avoid them?

Keywords

Publication bias · Selection bias · Reporting bias · Meta-analysis

7.1 Introduction

Systematic assessments of the best available information on the benefits and risks of medical interventions can help clinicians and public health officials make better decisions. These evaluations are based on meta-analysis, which is defined as 'a statistical analysis that combines or integrates the results of numerous separate clinical trials determined "combinable" by the analyst'. However, major randomised controlled studies eventually challenged the conclusions of several meta-analyses. Such inconsistencies have tarnished a technique that has been controversial from the start. Given the existence of publication bias and the many additional biases that can be introduced in the process of discovering, selecting, and merging papers, the occurrence of erroneous meta-analyses is not surprising. This chapter summarises the various types of bias involved in meta-analysis.

A. Kumar, *Meta-analysis in Clinical Research: Principles and Procedures*, https://doi.org/10.1007/978-981-99-2370-0_7

7.2 Publication Bias

Studies that only show a substantial difference are usually published, this means that some completed researches are not published and hence cannot be included in the meta-analysis. These could provide outcomes that differ from those found in published studies. As a result, if all studies had been evaluated, the meta-analysis could have yielded a different outcome. However, non-published papers are frequently given as abstracts for scientific meetings, we may do a more thorough analysis by incorporating the abstracts as well.

There are two types of publication bias. The conformity publication bias arises from the fact that if a common opinion on an issue has already been developed it will be simpler to publish studies that support that opinion, but discordant results will be less likely to be accepted. The inverse conformity publishing bias is the polar opposite of this 'editorial philosophy,' in which articles confirming previously established facts may be regarded redundant, whereas discordant discoveries may be accepted as innovative. The author, the trial sponsor, and the editorial policies of the journal to which the work has been submitted are all causes of publication bias.

Authors and journal editorial boards are typically less interested in publishing a paper that has poor results. Results that do not confirm the efficacy of a new treatment or intervention are usually unappealing to the study's sponsor.

A funnel plot can be used to detect publication bias qualitatively, while particular statistical tests such as the Begg and Egger tests can be used to analyse publication bias quantitatively (Thornton and Lee 2000; Mavridis and Salanti 2014; Duval and Tweedie 2000). Chapter 6 delves into the assessment of publication bias in further depth.

7.3 If the Search Is Internet Based (For Example, Medline)

7.3.1 Indexing Bias

This is owing to the internet database's inadequate indexing of studies. This mistake is more likely to occur when an article is brand new and has only recently been added to the database.

Checks are made in the second step, and mistakes are frequently repaired.

7.3.2 Search Bias

This is due to the database's inability to get all of the indexed papers. The rate of research discovered by an expert user using internet-based resources ranges from 32% to 80%, and it is much lower for inexperienced users. As a result, it is critical to pay close attention to the search approach. It is always a good idea to have many people run separate searches and then compare their findings.

There are various biases if the search is based on both internet-based databases and non-internet-based methods (literature review, etc.) that are mentioned below.

7.3.3 Reference Bias

This arises when some research is likely to be cited frequently while others are unlikely to be cited at all.

7.3.4 Multiple Publication Bias

This arises due to the same results being published in many journals.

7.3.5 Multiply Used Subjects Bias

This occurs as a result of the release of data from different subgroups of the same group of subjects.

7.3.5.1 Avoidance

All of these biases can be avoided by reading the papers carefully.

7.4 Selection Bias

7.4.1 Inclusion Criteria Bias

This can happen if the meta-inclusion analysis's criteria excludes relevant studies in some way. This error is difficult to detect, measure, and eliminate once the criteria have been chosen since it is an inclusion bias.

7.4.2 Selector Bias

If specific selection criteria are not stated from the start, the person conducting the search may use his or her own criteria, which may provide variable results. Defining exact selection criteria and evaluating research according to the criteria without considering their outcomes is an ideal strategy to eliminate bias.

7.5 Within Study Bias

It is possible that getting the data we need from the studies' results will be difficult once the papers have been chosen.

7.5.1 Bias Caused by the Meta-Analyst

Extractor bias: This occurs when the results are recorded incorrectly. This is a very regular occurrence that is tough to avoid. It could have a significant impact on the analyses' outcome. It is generally a good idea to do two separate readings and resolve any discrepancies.

Quality score bias: If the system for grading the quality of the papers are not clearly established, the reader's personal judgement may influence the analysis' conclusions. It is simple to avoid bias by not utilising a scoring system.

7.5.2 Bias Due to Inadequate Accuracy in Reporting the Results by the Authors of the Studies

Reporting bias: A trial planned to test numerous outcomes but only reporting the significant results is an example of this bias. This prejudice is inescapable and difficult to measure. To receive all the results, one can always contact the authors. Unreported data in a study always indicates the poor quality of studies study.

Recording error bias: It is owing to the authors' faults in interpreting the results obtained. It appears in around 1% of the data submitted. Because it is a random error, it is unimportant in a meta-analysis, especially if the number of individuals is large enough. It just adds a degree of uncertainty to the estimate (Egger and Smith 1998; Tierney and Stewart 2005; Williamson et al. 2005; Williamson and Gamble 2005; Carpenter et al. 2011).

7.6 Other Biases

Geographical bias: Because differences in the prevalence and incidence of diseases and risk factors may have significant effects on the results, it may not be advisable to aggregate the results of research conducted in various geographical locations. If this is the case, a meta-analysis of all the trials to produce an estimate of the mean effect may be unsuitable. Instead, a sensitivity analysis analysing subgroups of research from the same area should be undertaken to see whether all the results can be pooled together.

Follow-up time bias: If the outcome being studied has a time-dependent influence, the length of the follow-up is important. If the treatment is provided for different durations of time, studies utilising the same agent at the same dosage may yield different results. The evaluation of the mean effect might be skewed if the time-dependent effect is not taken into account, which is a prevalent problem in clinical settings.

7.7 Conclusion

There are a number of biases that can affect the outcome of a meta-analysis. It is critical to be aware of potential biases in order to avoid or at the very least mitigate their consequences.

References

Carpenter J, Rücker G, Schwarzer G (2011) Assessing the sensitivity of meta-analysis to selection bias: a multiple imputation approach. Biometrics 67(3):1066–1072

Duval S, Tweedie R (2000) Trim and fill: a simple funnel-plot–based method of testing and adjusting for publication bias in meta-analysis. Biometrics 56(2):455–463

Egger M, Smith GD (1998) Meta-analysis bias in location and selection of studies. BMJ 316(7124): 61–66

Mavridis D, Salanti G (2014) How to assess publication bias: funnel plot, trim-and-fill method and selection models. Evid Based Ment Health 17(1):30

Thornton A, Lee P (2000) Publication bias in meta-analysis: its causes and consequences. J Clin Epidemiol 53(2):207–216

Tierney JF, Stewart LA (2005) Investigating patient exclusion bias in meta-analysis. Int J Epidemiol 34(1):79–87

Williamson PR, Gamble C (2005) Identification and impact of outcome selection bias in meta-analysis. Stat Med 24(10):1547–1561

Williamson PR, Gamble C, Altman DG et al (2005) Outcome selection bias in meta-analysis. Stat Methods Med Res 14(5):515–524

Sensitivity and Subgroup Analysis

8

Abstract

Sensitivity and subgroup analysis are one of the important analyses in meta-analysis. A sensitivity analysis is a critical component of a meta-analysis since it tries to establish the robustness of the reported outcomes, whereas subgroup analysis provides a clear picture of the effect of interventions in different groups. However, these crucial aspects of meta-analyses are frequently ignored or underreported. This chapter provides a detailed discussion about sensitivity and subgroup analysis.

Keywords

Sensitivity · Subgroup analysis · Meta-analysis

8.1 Introduction

A meta-analysis synthesises the results of individual studies. The overall estimate is calculated depending upon the type of data. For example, for continuous data, we choose mean effect or standard mean difference, whereas, for dichotomous data, we usually calculate the relative risk or odd ratio. To continue with our previous examples in Chaps. 5 and 6 regarding the use of steroids in the reduction of mortality of COVID-19 patients, if we would like to find out which steroids are more effective and safer in reduction of deaths of COVID-19 patients, At what dosage? Which particular age group of patients? In a simpler way, this kind of analysis, we called it as a subgroup analysis. Another way is supposed if results of heterogenous trials are included in the meta-analysis so in that particular case it may be inappropriate to draw a valid conclusion. However, if trials were segregated into subgroups, it could result in a reduction of heterogeneity and would help to draw a valid conclusion.

A. Kumar, *Meta-analysis in Clinical Research: Principles and Procedures*,
https://doi.org/10.1007/978-981-99-2370-0_8

Therefore, subgroup analysis will provide us a more informative insight into treatment effectiveness that might not be obtained from the non-subgrouped analysis.

A sensitivity analysis is a critical component of a meta-analysis since it tries to establish the robustness of the reported outcomes to the assumptions used in the analysis (Bown and Sutton 2010; Copas and Shi 2000). Unfortunately, this crucial aspect of meta-analyses is frequently ignored or underreported. This chapter desribes the details of subgroup and sensitivity analysis.

8.2 Subgroup Analysis

Let us understand this through an example: one drug is equally effective in all age groups so it is appropriate to report the impact of that drug for the all-age groups. However, if the impact of that particular drug varies among different age groups. Then, the analysis must be done to take account of this difference. We may also check the long-term effects with short-term effects and so on. This is called a subgroup analysis. Therefore, the term 'subgroups' can refer to groupings based on a characteristic of the drugs as well as subjects. Subgroup analysis provides a clear picture of the effect of interventions in different groups. The decision to publish a summary effect across all subgroups is based on our objectives as well as the nature of the data.

Subgroup analysis compares the treatment effects of two or more variants of an intervention. While subgroup analyses are common in large-scale studies, they are not always possible in small-scale studies, which typically have a homogeneous population and just one variety of intervention. Meta-analysis can be used to get around this constraint. The researcher can compare the treatment impact in distinct subgroups using meta-analysis, even if the subgroups are from different trials.

Furthermore, even if there is little or no heterogeneity, authors may choose to provide results subgrouped by the patient group if it is clinically useful to evaluate therapy effectiveness for specific subgroups of patients. If subgroup analyses show that a treatment is more or less effective for certain subgroups of patients, the interpretation of these subgroup analyses might provide useful information about how the treatment should be employed in clinical practice. There are mainly three strategies for comparing the mean effect across subgroups. A Z-test can be used to compare the mean effect of two groups. We can use analysis of variance (adapted for use with subgroups) or the Q test of homogeneity to compare the mean effect in two or more groups.

8.3 How to Interpret Subgroup Analyses?

A statistically significant subgroup effect means that the covariate (trial or patient feature) examined in the subgroup analysis alters the treatment effect statistically significantly. The p-value from the test for subgroup differences should be considered when determining if a statistically significant subgroup difference was detected.

The difference between the pooled effect estimates for each subgroup is tested. A *p*-value of less than 0.1 for this test usually indicates a statistically significant subgroup effect.

Other details, however, may be useful to include when determining whether a statistically significant subgroup effect exists. It is important to note whether the subgroup effect is qualitative (each subgroup's treatment effects favour different treatments) or quantitative (each subgroup's treatment effects favour the same treatment but in different sizes), as well as the degree of heterogeneity (differences in treatment effects from individual trials in the meta-analysis) within each subgroup.

It may not be reasonable to draw conclusions regarding treatment effects within a subgroup if there is significant heterogeneity within that subgroup without further investigation of heterogeneity. If heterogeneity is discovered, the reviewers should examine whether the analysis should be presented in a way that is both appropriate and instructive. If the subgroup analysis was done to look for causes of heterogeneity, I recommend looking at the forest plots to see if there is less heterogeneity inside the subgroups than across all trials. If the subgroup analysis does not explain heterogeneity in any way, the authors may elect not to include it in the review. If the subgroup analysis was performed to provide estimates of treatment effects for clinically relevant subgroups of patients, the review authors may decide not to present the subgroup analysis if the results for each subgroup are meaningless due to heterogeneity within each subgroup or decide that it is important to show whether or not statistically significant subgroup differences exist for this covariate and thus present the analysis.

The review author must take into account the number of trials and participants in each subgroup. An assessment of heterogeneity is unlikely to provide useful conclusions until at least 10 trials are included in the meta-analysis, albeit even 10 trials may be insufficient if the covariate is unevenly distributed (i.e. if there is a limited amount of data for a particular subgroup).

Because of biological variability, therapy effects are likely to differ depending on the patient and/or trial factors like age, gender, and drug dosage. As a result, it would be remarkable if subgroup effects were not observed. However, if the differences in treatment effects are not significant enough to affect clinical judgements, it is not required to look into these subgroup effects any further. The greater the difference between the effect in a specific subgroup and the overall effect, the more significant the result. However, clinical specialists in the relevant field of research should be consulted to assess whether the subgroup analysis result is clinically significant.

The chance that a confounding factor is impacting the outcomes of the subgroup analysis, leading to inaccurate conclusions, should be considered by the review authors. If the effects of two factors on the treatment effect cannot be separated, they are said to be confounded. For example, consider a meta-analysis that includes trials comparing Treatment A to Treatment B. Because the intensity of Treatment A varies, some trials compare 'Intensive Treatment A' to 'Non-intensive Treatment A', while others compare 'Intensive Treatment A' to 'Non-intensive Treatment B'. Assume that a subgroup analysis was conducted to see if the intensity of Treatment

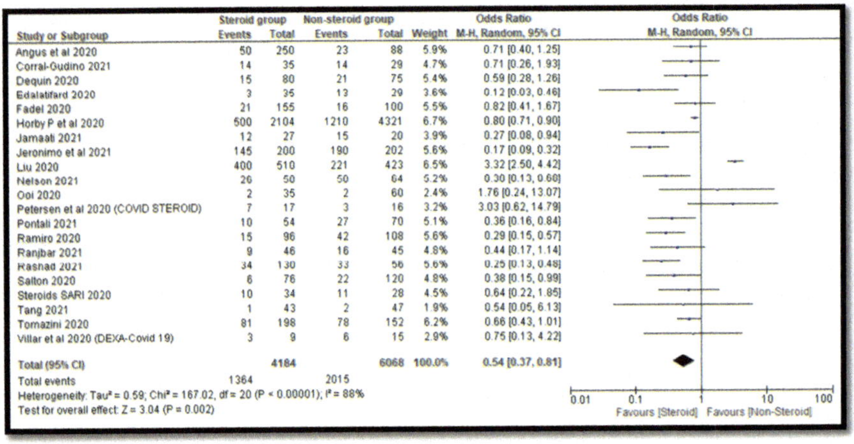

Study or Subgroup	Steroid group Events	Steroid group Total	Non-steroid group Events	Non-steroid group Total	Weight	Odds Ratio M-H, Random, 95% CI
Angus et al 2020	50	250	23	88	5.9%	0.71 [0.40, 1.25]
Corral-Gudino 2021	14	35	14	29	4.7%	0.71 [0.26, 1.93]
Dequin 2020	15	80	21	75	5.4%	0.59 [0.28, 1.26]
Edalatifard 2020	3	35	13	29	3.6%	0.12 [0.03, 0.46]
Fadel 2020	21	155	16	100	5.5%	0.82 [0.41, 1.67]
Horby P et al 2020	500	2104	1210	4321	6.7%	0.80 [0.71, 0.90]
Jamaati 2021	12	27	15	20	3.9%	0.27 [0.08, 0.94]
Jeronimo et al 2021	145	200	190	202	5.6%	0.17 [0.09, 0.32]
Liu 2020	400	510	221	423	6.5%	3.32 [2.50, 4.42]
Nelson 2021	26	50	50	64	5.2%	0.30 [0.13, 0.68]
Ooi 2020	2	35	2	60	2.4%	1.76 [0.24, 13.07]
Petersen et al 2020 (COVID STEROID)	7	17	3	16	3.2%	3.03 [0.62, 14.79]
Pontali 2021	10	54	27	70	5.1%	0.36 [0.16, 0.84]
Ramiro 2020	15	96	42	108	5.6%	0.29 [0.15, 0.57]
Ranjbar 2021	9	46	16	45	4.8%	0.44 [0.17, 1.14]
Rashad 2021	34	130	33	56	5.6%	0.25 [0.13, 0.48]
Salton 2020	6	76	22	120	4.8%	0.38 [0.15, 0.99]
Steroids SARI 2020	10	34	11	28	4.5%	0.64 [0.22, 1.85]
Tang 2021	1	43	2	47	1.9%	0.54 [0.05, 6.13]
Tomazini 2020	81	198	78	152	6.2%	0.66 [0.43, 1.01]
Villar et al 2020 (DEXA-Covid 19)	3	9	6	15	2.9%	0.75 [0.13, 4.22]
Total (95% CI)		**4184**		**6068**	**100.0%**	**0.54 [0.37, 0.81]**
Total events	1364		2015			

Heterogeneity: $Tau^2 = 0.59$; $Chi^2 = 167.02$, df = 20 ($P < 0.00001$); $I^2 = 88\%$

Test for overall effect: Z = 3.04 (P = 0.002)

Fig. 8.1 Role of steroids in reduction of mortality of COVID-19 patients

A affects the treatment impact, and that the subgroup analysis revealed a statistically significant treatment effect. It would be critical for the authors to think about whether there was another element that could be producing the subgroup differences. For example, that trials using a more intensive version of Treatment A enrolled individuals with more severe diseases, and that disease severity is the underlying impact modifier. If this were the case, concluding that the intensity of Treatment A alters the treatment effect would be erroneous. Authors of reviews should use their clinical experience to identify patient or trial features that may be confused with one another. To identify confounding factors, review authors may compute correlations between trial or patient characteristics.

To understand subgroup analysis in a simpler way, I have continued with our example related to the role of steroids in the reduction of mortality of COVID-19 patients (Thakur et al. 2022). The numbers included are imaginary.

Figure 8.1 shows the forest plot, which includes all studies related to the use of steroids in the reduction of mortality of COVID-19 patients. The hypothetical overall estimate was found to be 0.54 as an odds ratio with a confidence interval of [0.37, 0.81] which shows statistically significant results. Furthermore, if researchers would like to see the effect of individual steroids on the outcome, a subgroup analysis should be performed.

We have identified three steroids (dexamethasone, hydrocortisone, and methylprednisolone) that were used in COVID-19 patients and performed analysis. In this hypothetical example, we have found an odds ratio of 0.51 [0.30, 0.86] which indicates a significant role of dexamethasone in the reduction of mortality of COVID-19 patients, as shown in Fig. 8.2.

The results with hydrocortisone have shown non-significant [Odds ratio, 0.80 [0.43, 1.52]] effects in the reduction of deaths of COVID-19 patients as shown in Fig. 8.3.

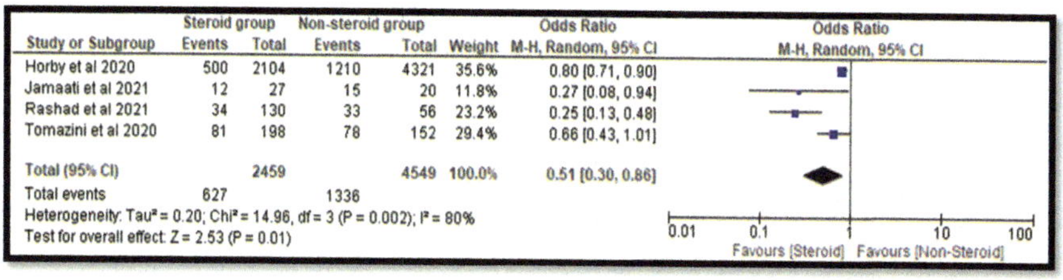

Fig. 8.2 Role of dexamethasone in reduction of mortality of COVID-19 patients

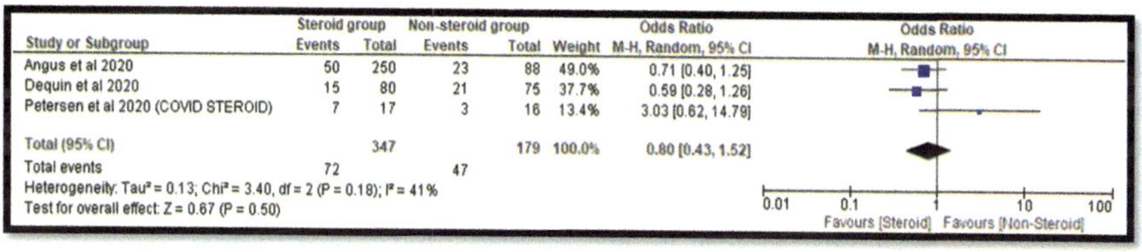

Fig. 8.3 Role of hydrocortisone in reduction of mortality of COVID-19 patients

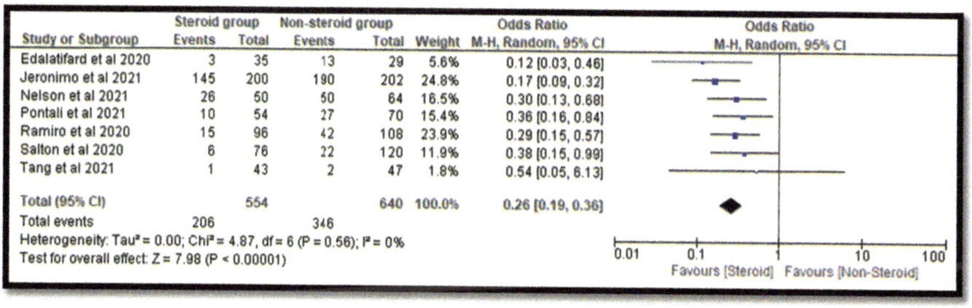

Fig. 8.4 Role of methylprednisolone in reduction of mortality of COVID-19 patients

The odds ratio in the case of methylprednisolone was found to be 0.26 [0.19, 0.36] which indicates significant reductions in the deaths of COVID-19 patients as shown in Fig. 8.4.

8.4 Sensitivity Analysis

The outlier affects the results of the meta-analysis in a significant way. Recently, our lab performed a meta-analysis to find out the role of aspirin in the reduction of deaths of COVID-19 patients and found a significant reduction in the aspirin group as compared to the non-aspirin group. However, when we excluded studies with very high or low sample sizes, we found a non-significant effect. Therefore, sensitivity analysis can be done in a systematic way. First, an overall estimate can be calculated by including all studies, and second, only including those that are definitely known to be eligible. A sensitivity analysis asks the question, 'Are the findings robust to the decisions made in the process of obtaining them?' (Elvik 2005).

There are a number of factors that should be considered for sensitivity analysis. For example, if not all participants meet an age range as per the inclusion criteria of the study, should the study be included? How many different doses should be included? What are the requirements for defining a comparator group? What time period (or range of time periods) is acceptable for inclusion? Should studies be excluded based on other parts of methodological standards, or should be blinded and unblinded outcome assessment be included? When and how should standard deviations be estimated when they are missing? Should final values or change scores be used in analyses? What should be the cut-off point for dividing short ordinal scales into two groups? When trial analyses have not been adjusted for clustering, what values of the intraclass correlation coefficient should be used? When the within-subject correlation coefficient is not available in primary reports, what values should be used? To make intention-to-treat studies easier, what assumptions should be made about missing outcomes? Should treatment effect estimates be adjusted or unadjusted?

It would be better if the author pre-specified regarding sensitivity analysis in the study protocol. However, many issues the author came to know during the process of meta-analysis. When sensitivity analyses show that the overall result and conclusions are unaffected by the various decisions that could be made during the review process, the review results can be viewed with greater confidence. When sensitivity analyses reveal specific decisions or missing data that have a significant impact on the review's conclusions, further resources might be allocated to resolving doubts and obtaining additional information, for as by contacting trial authors and obtaining individual patient data. If this is not possible, the results must be regarded with caution. Such findings may lead to suggestions for additional studies and inquiries. The best way to report sensitivity analyses in a systematic review is to create a summary table.

To continue with our previous example of the role of steroids in the reduction of deaths of COVID-19 patients, (Fig. 8.5).

The sensitivity analysis was done to check the effect of outliers. In the current analysis, we have found two studies, i.e. Horby et al. (2021) and Liu et al. (2020) with a high sample size, whereas one study, i.e. Villar et al. (2020), has a low sample size. The pooled OR was found to be 0.45 [0.34, 0.60] after the exclusion of both studies with a high sample size, which also indicates a significant decrease in the deaths of COVID-19 patients in the steroidal group as compared to the non-steroidal group (Fig. 8.6).

We have also done analysis again after exclusion of the low sample size, Villar et al. (2020) and the odd ratio was found to be 0.54 [0.36, 0.80] which also indicates a statistically significant reduction in deaths of COVID-19 patients in the steroidal group as compared to the non-steroidal group as shown in Fig. 8.7.

Finally, analysis was also done to check the effect of both low and high sample size studies on the conclusion of the study and also found statistically significant results as shown in Fig. 8.8.

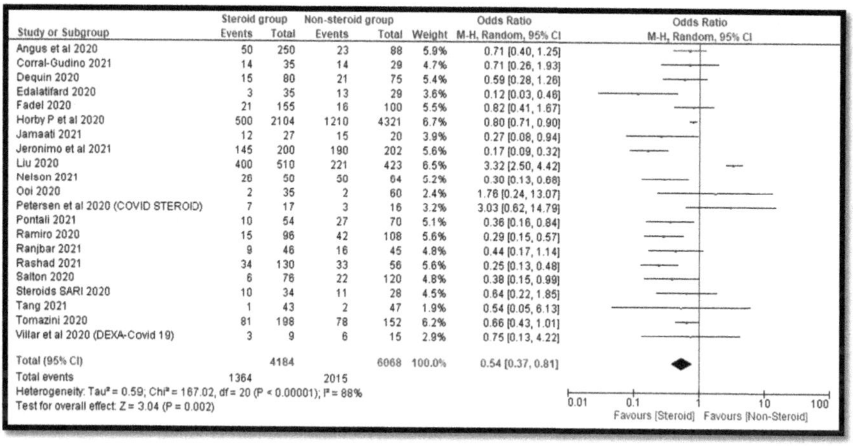

Study or Subgroup	Steroid group Events	Steroid group Total	Non-steroid group Events	Non-steroid group Total	Weight	Odds Ratio M-H, Random, 95% CI
Angus et al 2020	50	250	23	88	5.9%	0.71 [0.40, 1.25]
Corral-Gudino 2021	14	35	14	29	4.7%	0.71 [0.26, 1.93]
Dequin 2020	15	80	21	75	5.4%	0.59 [0.28, 1.26]
Edalatifard 2020	3	35	13	29	3.6%	0.12 [0.03, 0.46]
Fadel 2020	21	155	16	100	5.5%	0.82 [0.41, 1.67]
Horby P et al 2020	500	2104	1210	4321	6.7%	0.80 [0.71, 0.90]
Jamaati 2021	12	27	15	20	3.9%	0.27 [0.08, 0.94]
Jeronimo et al 2021	145	200	190	202	5.6%	0.17 [0.09, 0.32]
Liu 2020	400	510	221	423	6.5%	3.32 [2.50, 4.42]
Nelson 2021	26	50	50	64	5.2%	0.30 [0.13, 0.68]
Ooi 2020	2	35	2	60	2.4%	1.76 [0.24, 13.07]
Petersen et al 2020 (COVID STEROID)	7	17	3	16	3.2%	3.03 [0.62, 14.79]
Pontali 2021	10	54	27	70	5.1%	0.36 [0.16, 0.84]
Ramiro 2020	15	96	42	108	5.6%	0.29 [0.15, 0.57]
Ranjbar 2021	9	46	16	45	4.8%	0.44 [0.17, 1.14]
Rashad 2021	34	130	33	56	5.6%	0.25 [0.13, 0.48]
Salton 2020	6	76	22	120	4.8%	0.38 [0.15, 0.99]
Steroids SARI 2020	10	34	11	28	4.5%	0.64 [0.22, 1.85]
Tang 2021	1	43	2	47	1.9%	0.54 [0.05, 6.13]
Tomazini 2020	81	198	78	152	6.2%	0.66 [0.43, 1.01]
Villar et al 2020 (DEXA-Covid 19)	3	9	6	15	2.9%	0.75 [0.13, 4.22]
Total (95% CI)		4184		6068	100.0%	0.54 [0.37, 0.81]
Total events	1364		2015			

Heterogeneity: Tau² = 0.59; Chi² = 167.02, df = 20 (P < 0.00001); I² = 88%
Test for overall effect: Z = 3.04 (P = 0.002)

Fig. 8.5 Role of steroids in reduction of deaths of COVID-19 patients

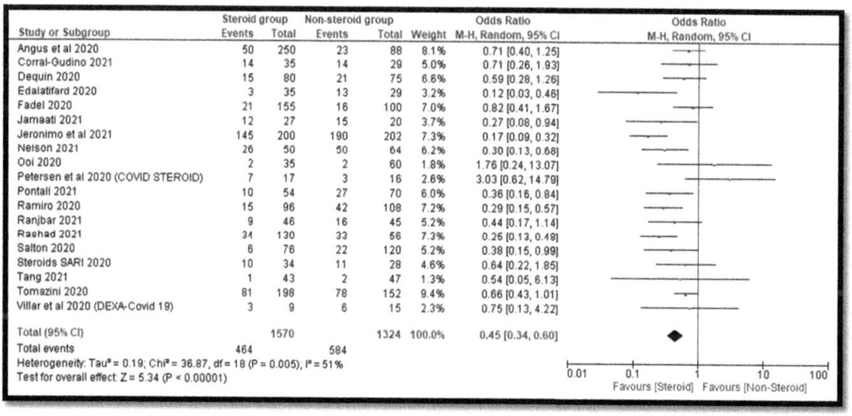

Study or Subgroup	Steroid group Events	Steroid group Total	Non-steroid group Events	Non-steroid group Total	Weight	Odds Ratio M-H, Random, 95% CI
Angus et al 2020	50	250	23	88	8.1%	0.71 [0.40, 1.25]
Corral-Gudino 2021	14	35	14	29	5.0%	0.71 [0.26, 1.93]
Dequin 2020	15	80	21	75	6.6%	0.59 [0.28, 1.26]
Edalatifard 2020	3	35	13	29	3.2%	0.12 [0.03, 0.46]
Fadel 2020	21	155	16	100	7.0%	0.82 [0.41, 1.67]
Jamaati 2021	12	27	15	20	3.7%	0.27 [0.08, 0.94]
Jeronimo et al 2021	145	200	190	202	7.3%	0.17 [0.09, 0.32]
Nelson 2021	26	50	50	64	6.2%	0.30 [0.13, 0.68]
Ooi 2020	2	35	2	60	1.8%	1.76 [0.24, 13.07]
Petersen et al 2020 (COVID STEROID)	7	17	3	16	2.6%	3.03 [0.62, 14.79]
Pontali 2021	10	54	27	70	6.0%	0.36 [0.16, 0.84]
Ramiro 2020	15	96	42	108	7.2%	0.29 [0.15, 0.57]
Ranjbar 2021	9	46	16	45	5.2%	0.44 [0.17, 1.14]
Rashad 2021	34	130	33	56	7.3%	0.25 [0.13, 0.48]
Salton 2020	6	76	22	120	5.2%	0.38 [0.15, 0.99]
Steroids SARI 2020	10	34	11	28	4.6%	0.64 [0.22, 1.85]
Tang 2021	1	43	2	47	1.3%	0.54 [0.05, 6.13]
Tomazini 2020	81	198	78	152	9.4%	0.66 [0.43, 1.01]
Villar et al 2020 (DEXA-Covid 19)	3	9	6	15	2.3%	0.75 [0.13, 4.22]
Total (95% CI)		1570		1324	100.0%	0.45 [0.34, 0.60]
Total events	464		584			

Heterogeneity: Tau² = 0.19; Chi² = 36.87, df = 18 (P = 0.005); I² = 51%
Test for overall effect: Z = 5.34 (P < 0.00001)

Fig. 8.6 Role of steroids in reduction of deaths of COVID-19 patients after exclusion of high sample size (Horby et al. (2021) and Liu et al. (2020))

In conclusion, there is no significant effect of outliers on the conclusion of the study.

In another way, to understand sensitivity analysis, I have added another hypothetical study and created a forest plot as shown in Fig. 8.9. In this particular example, you can see mainly two studies with very high sample sizes. Therefore, we have also performed analysis after the exclusion of these two studies.

The odd ratios (OR) were found to be 0.70 [0.48, 1.03] which indicates a non-significant reduction in deaths of COVID-19 patients in the steroidal group as compared to the non-steroidal group as shown in Fig. 8.10. Therefore, there is an effect of outliers on the conclusion of the study.

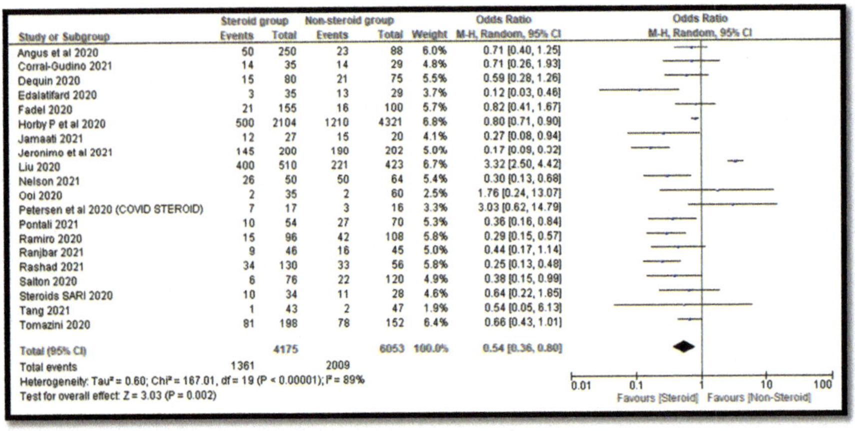

Fig. 8.7 Role of steroids in reduction of deaths of COVID-19 patients after exclusion of low sample size (Villar et al. 2020)

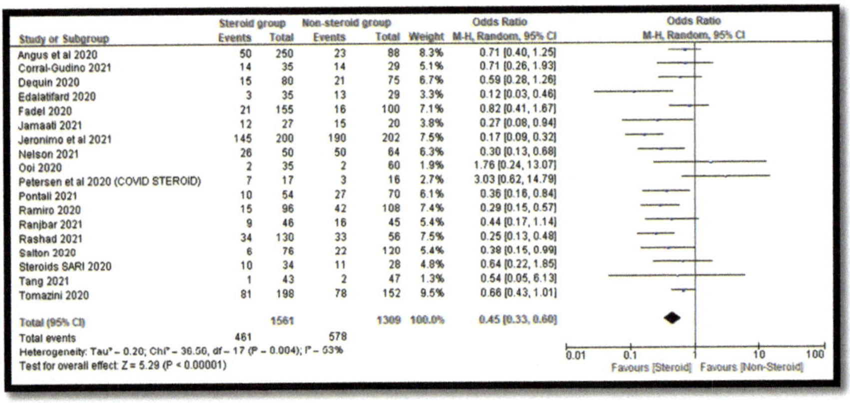

Fig. 8.8 Role of steroids in reduction of deaths of COVID-19 patients after exclusion of all high and low sample size studies

8.5 Sensitivity Analyses Vs. Subgroup Analysis

Subgroup analysis and sensitivity analyses are sometimes misconstrued. Despite the fact that certain sensitivity analyses limit the analysis to a subset of the entire number of studies, the two methodologies differ in two ways. First, whereas subgroup analyses give estimates for each subgroup, sensitivity analyses do not attempt to quantify the effect of the intervention in the group of studies eliminated from the analysis. Second, informal comparisons are done between different ways of estimating the same object in sensitivity analyses, whereas formal statistical

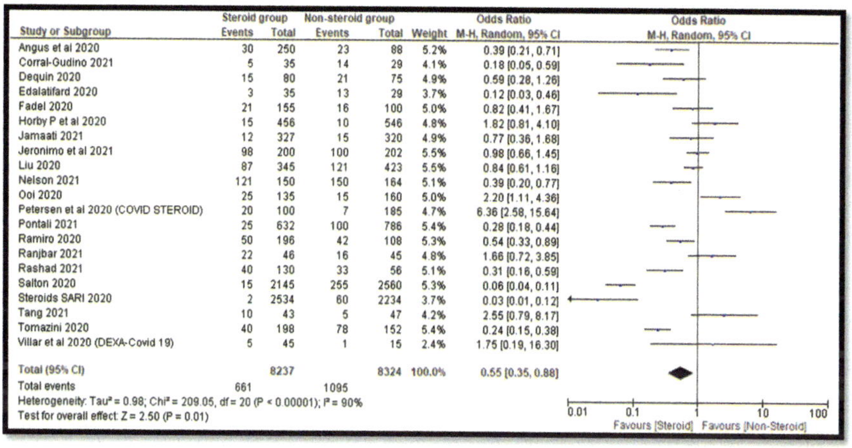

Fig. 8.9 Forest plot of hypothetical studies

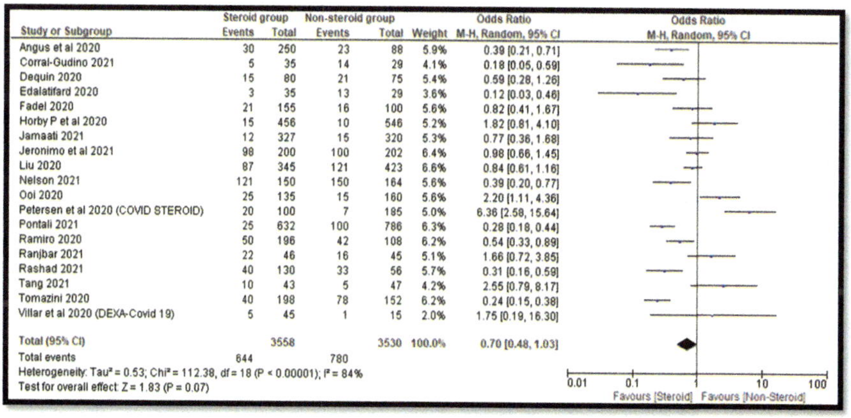

Fig. 8.10 Forest plot after exclusion of Salton 2020 and steroids SARI 2020

comparisons are made among subgroups in subgroup analysis (Richardson et al. 2019).

8.6 Conclusion

In conclusion, subgroup analysis is helpful to provide the effect of individual parameters on the outcome as well as explain the reason for heterogeneity, whereas sensitivity analysis helps us to provide the effect of outliers on the outcome.

References

Bown MJ, Sutton AJ (2010) Quality control in systematic reviews and meta-analyses. Eur J Vasc Endovasc Surg 40(5):669–677

Copas J, Shi JQ (2000) Meta-analysis, funnel plots and sensitivity analysis. Biostatistics 1(3): 247–262

Elvik R (2005) Can we trust the results of meta-analyses?: a systematic approach to sensitivity analysis in meta-analyses. Transp Res Rec 1908(1):221–229

Horby P, Lim WS, Emberson JR, Mafham M, Bell JL, Linsell L, Staplin N, Brightling C, Ustianowski A, Elmahi E, Prudon B, Green C, Felton T, Chadwick D, Rege K, Fegan C, Chappell LC, Faust SN, Jaki T, Jeffery K, Montgomery A, Rowan K, Juszczak E, Baillie JK, Haynes R, Landray MJ (2021) Dexamethasone in hospitalized patients with covid-19. N Engl J Med 384:693–704

Liu J, Zhang S, Dong X, Li Z, Xu Q, Feng H, Cai J, Huang S, Guo J, Zhang L, Chen Y, Zhu W, Du H, Liu Y, Wang T, Chen L, Wen Z, Annane D, Qu J, Chen D (2020) Corticosteroid treatment in severe COVID-19 patients with acute respiratory distress syndrome. J Clin Invest 130:6417–6428

Richardson M, Garner P, Donegan S (2019) Interpretation of subgroup analyses in systematic reviews: a tutorial. Clin Epidemiol Glob Health 7(2):192–198

Thakur M, Datusalia AK, Kumar A (2022) Use of steroids in COVID-19 patients: a meta-analysis. Eur J Pharmacol 914:174579

Villar J, Anon JM, Ferrando C, Aguilar G, Munoz T, Ferreres J, Ambros A, Aldecoa C, Suarez-Sipmann F, Thorpe KE, Jüni P, Slutsky AS, Ferrando C, Mellado-Artigas R, Fernandez J, Hernandez M, Castella M, Castro P, Badia JR, Aguilar G, Carbonell JA, Badenes R, Tornero C, Ferreres J, Blasco ML, Carbonell N, Serrano A, Juan M, Gomez-Herreras JI, Lopez ML, Ambros A, Martín C, DelCampo R, Puig-Bernabeu J, Ferrer C, De Andres J, Munoz T, Serna-Grande P, Tamayo G, Martínez-Ruíz A, Bilbao-Villasante I, Villar J, Fernandez RL, Calvo CP, Vidal A, Anon JM, Figueira JC, Asensio MJ, Maseda E, Suarez-Sipmann F, Ramasco F, Varela-Duran M, Díaz-Parada P, Trenado-Alvarez J, Fernandez MM, Aldecoa C, Rico-Feijoo J, Fernandez L, Sanchez-Ballesteros J, Blanco-Schweizer P, Martínez D, Soler JA, Slutsky AS, Jüni P, Thorpe KE, Thomas R, Wysocki K, De Verno P, Lakhanpal G, JuandoPrats C (2020) Efficacy of dexamethasone treatment for patients with the acuterespiratory distress syndrome caused by COVID-19: study protocol for a randomized controlled superiority trial. Trials 21:717

Meta-Regression

9

Abstract

Meta-regression is a powerful tool for exploratory analyses of heterogeneity using regression methods. It is an extension of a standard meta-analysis that investigates the extent to which statistical heterogeneity between results of multiple studies can be related to one or more characteristics of the studies. It uses regression analysis to combine, compare, and synthesise research findings from multiple studies. This chapter provides a detailed discussion on meta-regression.

Keywords

Meta-regression · Heterogeneity · Regression methods

9.1 Introduction

Usually, we stop our analysis after calculating the overall estimate measure. However, it is very important to do further analysis if heterogeneity among studies is found to be very high. Meta-regression is a statistical method used in meta-analysis to explain heterogeneity in terms of characteristics of the studies or study populations, etc. Subgroup analysis and sensitivity analyses are sometimes misconstrued. Despite the fact that certain sensitivity analyses limit the analysis to a subset of the entire number of studies, the two methodologies differ in two ways whereas subgroup analyses give estimates for each subgroup. Sensitivity analyses do not attempt to quantify the effect of the intervention in the group of studies eliminated from the analysis. Second, informal comparisons are made between different ways of estimating the same object in sensitivity analyses, whereas formal statistical comparisons are made among subgroups in subgroup analysis.

A. Kumar, *Meta-analysis in Clinical Research: Principles and Procedures*,
https://doi.org/10.1007/978-981-99-2370-0_9

Meta-regression is a powerful tool for exploratory analyses of heterogeneity using regression methods. It can detect interactions of treatment with study-level factors when treatment effects are heterogeneous (Schmid et al. 2004). It is used to create a model describing the linear relationship between (both continuous and categorical) study-level covariates and the effect size. In other words, meta-analysis regression, or meta-regression, is an extension of a standard meta-analysis that investigates the extent to which statistical heterogeneity between results of multiple studies can be related to one or more characteristics of the studies (Thompson and Higgins 2002). This chapter provides a detailed discussion on meta-regression.

9.2 Meta-Regression Approaches

There are several meta-regression methods. The first method is a fixed effects method based on logistic regression. A weighted logistic regression of the 2 k cases per study is fitted in this technique, where k is the number of study arms and weight is the number of patients who have or do not have the outcome. Interactions with treatment can be fit, and covariates can be a study or arm level. Random effects meta-regression is the second method. The log odds ratio is usually regressed on an intercept as well as study-level factors.

The phrase 'random effects' refers to the inclusion of a random study effect in the regression model to account for variation between studies. This approach simplifies to the typical DerSimonian and Laird random effects estimate of the pooled odds ratio in the simplest case, where just an intercept factor is included. Control rate meta-regression is the third method. In this case, the outcome rate (e.g. mortality rate) in the control group is the only covariate. The control rate is a substitute for covariate differences between studies, according to the theory. Bayesian hierarchical modelling is the fourth general strategy (Luebke and Brunkwall 2015; Thompson et al. 2002; Bellavance et al. 2009).

9.3 Software

All available software are not useful to perform meta-regression. The most commonly used software to perform meta-regression are R package, MetaXL 5.3, STATISTICA, etc. The details regarding individual software are presented in Chap. 12. A brief summary is presented below.

9.3.1 MetaXL 5.3

MetaXL is a Microsoft Excel for Windows add-in for meta-analysis. It supports all major meta-analysis methods, as well as the inverse variance heterogeneity and quality effects models, which are unique. Starting with v4.0, it also includes a powerful yet simple method for performing network meta-analysis.

9.3.2 Meta-Regression in R

The meta package of R includes a function called metareg that allows us to perform a meta-regression. As input, the metareg function only requires a {meta} meta-analysis object and the name of a covariate. Using the bubble function, the meta package allows us to visualise a meta-regression. This generates a bubble plot that displays the estimated regression slope as well as the effect size for each study. The bubbles are different sizes to represent the weight of a study, with a larger size representing a higher weight.

9.3.3 Statistica

STATISTICA is also useful to perform meta-regression. The typical procedure in STATISTICA is mentioned below:

1. Navigate to the Regression.sta data set, which contains study-level predictors.
2. Next, in the Additional variables group, click Edit.
3. In the Additional variables dialogue, add the variables Continuous, 'Measurement', and 'Categorical', as well as the variable 'Group'. Press the OK button.
4. In the Manage additional variables dialogue, two new columns, Measurement, and Group, will be added to the grid. You can manually enter predictor values or import data from a spreadsheet. Click the Import button, navigate to and select the Regression.sta spreadsheet, and then click OK.
5. Studies are distinguished by their titles. Then, to return to the Meta-analysis and Meta-regression dialogue, click the OK button.
6. Select variables for analysis on the Regression tab. In the Predictors group, select 'log OR (log odds ratio)' as the 'Effect measure-dependent variable' and click Select.
7. Next, in the Meta-regression predictors dialogue, select continuous variable Measurement and click OK.
8. On the Regression tab, a new column, Measurement, will be added to the grid. Click the Report button in the Meta-regression group to create a meta-regression model. Depending on which option is selected in the 'Model assumptions' group on the Settings tab, the programme will calculate the fixed effect or random effects model.
9. The coefficient for the Measurement variable will be calculated.
10. A bubble plot can be used to display the results of the meta-regression model. To make one, go to the Meta-regression group and select 'Bubble plot'. The bubble plot displays the value of the predictor Measurement on the horizontal axis and the effect measure 'log OR' on the vertical axis for each study. Each bubble's area represents the weight of the corresponding study in the meta-regression model.
11. To include a categorical variable in the model, click 'Select' in the Predictors group. Next, select the additional categorical variable 'Group' in the

Meta-regression predictors dialogue, select a hidden level (e.g. A) and click OK. Now you can see codes of groups in the grid on the Regression tab.
12. To select the type of categorical predictors coding, use the 'Categorical variable' combo box. The options enable you to use 'Reference coding', 'Sigma restricted coding', or to treat the groups as 'Strata'. Select Reference coding. Next, click the Bubble plot button. When you select more than one predictor, the program will show the Variable values dialogue in which you can set the values of variables not displayed on the two-dimensional plot. Use the default settings and click OK to create a categorised bubble plot.

9.4 Limitations

Despite of various advantages of meta-regression, there are some of the limitations that can impair the ability of the model to make valid inferences. Usually, a sufficient sample size is required to perform meta-regression which is not always available and low sample sizes are more prone to bias. Further, published studies may not always measure or appropriately report the information on covariates which results in the inability to properly adjust the confounders.

9.5 Applications

Researchers should understand meta-regression is not always required in every meta-analysis. Sometimes, meta-analysis results are sufficient to derive a valid conclusion. Meta-regression is required to perform if heterogeneity among studies was found to be high. Meta-regression can also serve to implement network meta-analysis. In this case, the model will use information about a parameter calculated in individual study groups defined by levels of exposure(s). For instance, instead of using the study-specific log OR as an outcome, it is possible to model the log odds in each contrasted group (placebo, exposure1, exposure2, etc.). Then, the exposure specification can be used in meta-regression as a moderator of the log odds to obtain a log OR.

9.6 Conclusion

For exploratory heterogeneity analysis, meta-regression is a useful tool. Meta-regression can be done using either fixed effects or a random effects model, but the latter is usually preferable. Meta-regression may also aid in better understanding the sources of variability in meta-analysis and provide a deeper summary of published data than a single number or point estimate. Furthermore, meta-regression is easily linked to the increasingly common multilevel modelling techniques. It must, however, be carried out with extreme caution, as it is prone to errors, poor methodological execution, and misinterpretations.

References

Bellavance F, Georges D, Martin L (2009) The value of a statistical life: a meta-analysis with a mixed effects regression model. J Health Econ 28(2):444–464

Luebke T, Brunkwall J (2015) Meta- analysis and meta-regression analysis of the associations between sex and the operative outcomes of carotid endarterectomy. BMC Cardiovasc Disord 15:32

Schmid CH, Stark PC, Berlin JA et al (2004) Meta-regression detected associations between heterogeneous treatment effects and study-level, but not patient-level, factors. J Clin Epidemiol 57(7):683–697

Thompson SG, Higgins JPT (2002) How should meta-regression analyses be undertaken and interpreted? Stat Med 21:1559–1573

Thompson SG, Simon G, Higgins Julian PT (2002) How should meta-regression analyses be undertaken and interpreted? Stat Med 21(11):1559–1573

Plots 10

Abstract

Forest and funnel plots are the most commonly used plots in every type of meta-analysis. The forest plot tells us about the estimate measure of individual studies as well as the overall estimate measure along with confidence interval. In addition, the forest plot also tells us about heterogeneity among studies by calculating Tau^2, Chi^2, and I^2 values as well as weightage provided to individual studies. The qualitative assessment of publication bias is done using a funnel plot. In this chapter, I have discussed the forest plot and funnel plot along with suitable examples for a better understanding of the readers.

Keywords

Forest plot · Funnel plot · Meta-analysis

10.1 Introduction

The most commonly used plots in meta-analysis are forest and funnel plots. The forest plot tells us about the overall estimate measure, whereas the funnel plot tells us about publication bias. This chapter provides detailed information about various plots used in meta-analysis.

10.2 How to Read a Forest Plot?

Figure 10.1 represents the impact of steroids on COVID-19 patients. This meta-analysis is published in the European Journal of Pharmacology by Thakur et al. (2022). The goal of the presentation is to understand the various elements of a forest plot, such as effect measures, weightage to study, overall effect measures, and so on.

A. Kumar, *Meta-analysis in Clinical Research: Principles and Procedures*,
https://doi.org/10.1007/978-981-99-2370-0_10

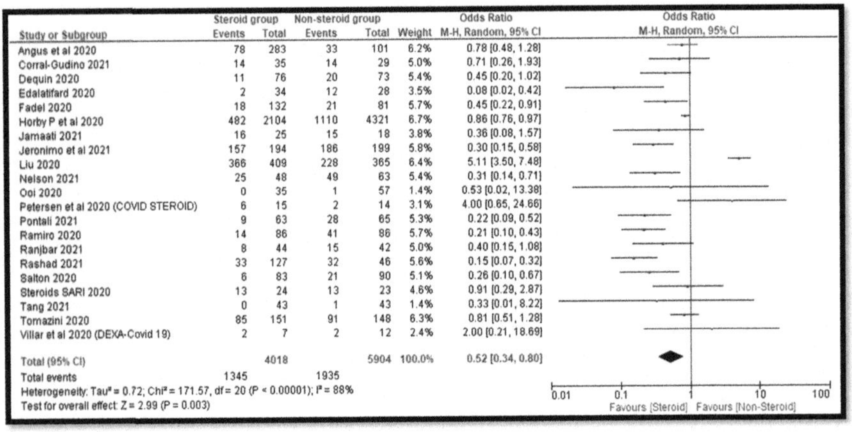

Fig. 10.1 Pooled analysis results using a random effect model (forest plot)

The individual rows on this plot represent the individual studies. The name of the study is indicated on the left side. The total number of patients in steroidal groups and non-steroidal groups is presented along with events in both groups. The next column represents the weightage of an individual study, followed by the effect size in terms of odds ratio along with the confidence interval.

The effect size represents the magnitude of the treatment effect. We calculate the impact size for each study and use it to analyse the consistency of the effect across studies and compute a summary effect. The effect size can be calculated as an odd ratio, relative risk, risk difference, mean difference, standard mean difference, etc., depending upon the collected data. The effect size demonstrates the impact of medicines in the treatment of diseases, the impact of treatment on the risk of infection, the prevalence of various problems in patients, the effect of treatment on laboratory parameters, etc. In this particular example, we would like to see the effect of steroids in the mortality of COVID-19 patients, and the effect size is calculated as an odd ratio. The effect size also represents the relationship between variables, such as the effect of treatment in different age groups, sex, etc. The effect size in this particular example is an odd ratio, and an odd ratio of 1.0 means that the effect of steroids in the reduction of the mortality of COVID-19 patients as same in non-steroidal groups. The odd ratio of less than 1.0 means the non-steroidal group has a significant effect in reduction of mortality of COVID-19 patients, whereas the odd ratio of less than 1.0 means steroids have a significant effect in the reduction of mortality of COVID-19 patients. The square in the middle of the horizontal lines in the forest plot represents the effect size of the individual studies. The square can be in both directions and can represent the magnitude of effect on both sides. In this particular example, the effect size of most of the studies falls to the left of the centre which indicates a benefit in the steroidal group. However, three studies represent the effect size in the right of the centre which indicates benefits in the non-steroidal group. The effect size based on ratios like odd ratio indicates no difference between groups. However, the effect size is based on differences like mean difference, a difference of 0.0 represents no difference between groups.

The effect size is always represented along with a confidence interval. The horizontal lines in the forest plot represent a confidence interval. The narrow line indicates higher precision and more weightage is provided for that particular study. The size of the solid squares used to represent each of the studies varies, with the size of each square representing the weight allocated to the relevant study when the summary effect is computed.

As one might imagine, there is a link between the precision of a study and its weight in the analysis. Studies with a high degree of precision are given more weight, whereas those with a low degree of precision are given less. We can think of the studies as being weighted by sample size because accuracy is mostly determined by sample size. While precision is one of the factors considered when assigning weights, there are often others.

10.2.1 *p*-Values

The *p*-value for a null hypothesis test is shown for each study. The *p*-value and the confidence interval must be in agreement, such that the *p*-value will be less than 0.05 if and only if the 95% confidence interval excludes the null value. As a result, we can quickly find statistically significant studies by scanning the confidence intervals.

The bottom line of the diagram depicts the summary effect. The summary odd ratio, in this case, is 0.52, showing that deaths in the steroidal group are lower than those in the non-steroidal group. The weighted mean of the individual effects is what the overall effect is. The mechanism utilised to allocate the weights (and thus the meaning of the overall effect) is, however, dependent on our assumptions about the effect size distribution from which the studies were sampled. We assume that all studies in the analysis have the same genuine effect size in the fixed effect model, and the summary effect is our estimate of this common effect size. We assume that the true impact size varies from study to study under the random effects model, and the summary effect is our estimate of the mean of the range of effect sizes.

10.2.2 Diamond Shape

A diamond represents the overall effect. The size of the influence is represented by the diamond's placement, while the precision of the estimate is shown by its width. The diamond in this example is centred at 0.52, and it spans 0.34–0.80. The precision refers to the precision with which the summary effect as an estimate of the genuine effect is calculated.

10.2.2.1 *p*-Value

The summary effect has a p-value of 0.003. This p-value represents the magnitude of the overall effect size as well as the amount of data used to make the estimate. It is worth noting that the p-value for the overall effect is far higher than that of any single study.

To conduct a meta-analysis, we first calculate the effect size and variance for each study, then calculate the weighted mean of these effect sizes. We give more weight to the more detailed studies when computing the weighted mean, but the procedures for doing so are based on our assumptions about the distribution of genuine effects.

The accuracy, which is commonly expressed as a standard error or confidence interval, indicates how certain we may be about the effect magnitude. The difference between reporting an effect size of 0.50 plus/minus 0.10 and reporting an effect size of 0.50 plus/minus 0.50 is significant. A standard error or confidence interval (in the same metric as the effect size itself) or variance can be used to illustrate the precision with which we estimate an effect size (in a squared metric). The precision is mostly determined by sample size, with larger research producing more precise impact size estimates.

The study design also affects precision, with matched groups producing more exact estimates (as opposed to independent groups) and clustered groups producing less accurate estimates. Studies that produce more precise impact size estimates have more information and are given more weight in the meta-analysis (Andrade 2020; Alavi et al. 2021; Sedgwick 2015; Stephenson 2017).

10.3 Funnel Plot

The symmetrical plot in the absence of bias (open circles indicate smaller studies showing no beneficial effects).

The asymmetrical plot in the presence of publication bias (smaller studies showing no beneficial effects are missing).

The asymmetrical plot in the presence of bias due to the low methodological quality of smaller studies (open circles indicate small studies of inadequate quality whose results are biased towards larger beneficial effects) (Fig. 10.2).

Plots the effect size against the sample size of the study. To study a funnel plot, look at its LOWER LEFT corner, that is where negative or null studies are located If EMPTY, this indicates 'PUBLICATION BIAS'. Note that here, the plot fits in a funnel, and that the left corner is not all that empty, but we cannot rule out publication bias (Egger et al. 1997; Sterne and Egger 2001).

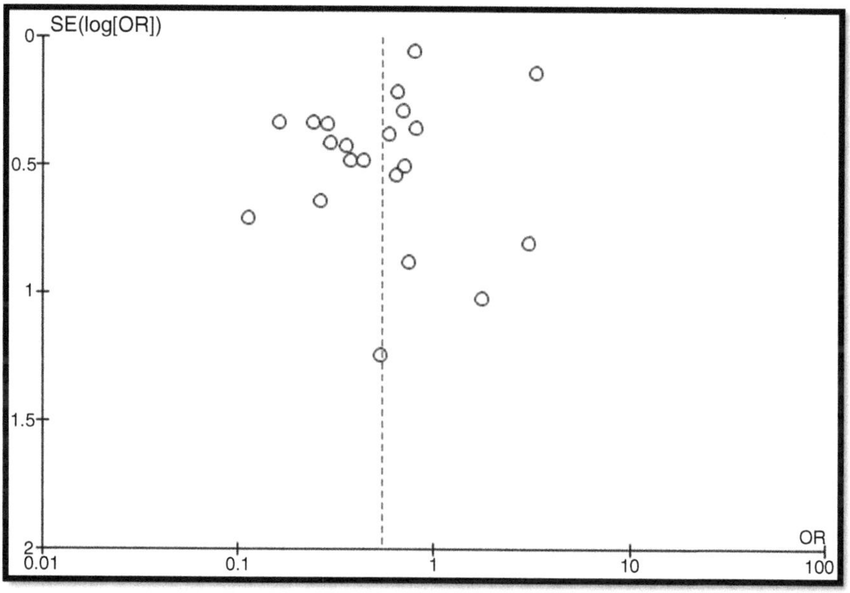

Fig. 10.2 Funnel plot

10.4 Conclusion

In conclusion, forest and funnel plots are most commonly used in every type of meta-analysis that provides information related to the overall estimate measure along with a confidence interval as well as information related to publication bias.

References

Alavi M, Hunt GE, Visentin DC et al (2021) Seeing the forest for the trees: how to interpret a meta-analysis forest plot. J Adv Nurs 77(3):1097–1101

Andrade C (2020) Understanding the basics of meta-analysis and how to read a forest plot: as simple as it gets. J Clin Psychiatry 81(5):20f13698

Egger M, Davey Smith G, Schneider M et al (1997) Bias in meta-analysis detected by a simple, graphical test. BMJ 315(7109):629–634

Sedgwick P (2015) How to read a forest plot in a meta-analysis. BMJ 351:h4028

Stephenson J (2017) Explaining the forest plot in meta-analyses. J Wound Care 26(11):611–612

Sterne JA, Egger M (2001) Funnel plots for detecting bias in meta-analysis: guidelines on choice of axis. J Clin Epidemiol 54(10):1046–1055

Thakur M, Datusalia AK, Kumar A (2022) Use of steroids in COVID-19 patients: a meta-analysis. Eur J Pharmacol 914:174579

Network Meta-Analysis

11

Abstract

Network meta-analysis compares multiple interventions simultaneously by analysing studies, making different comparisons in the same analysis, also known as mixed-treatment comparison or multiple treatment comparison. It is therefore a key tool for evidence-based medicine. This type of meta-analysis can be used to compare two interventions that have not been studied together in the same trial. It compares several treatments, utilising direct comparisons of interventions within randomised controlled trials as well as indirect comparisons across trials based on a common comparator. This chapter provides a detailed discussion of network meta-analysis.

Keywords

Network · Models · Regression · Network-meta-analysis

11.1 Introduction

In a typical meta-analysis of interventional studies, we compare the efficacy or safety of exactly two treatments that have been directly compared in clinical trials by calculating the overall estimate measure of all individual studies of one specific treatment. This will help to find out whether the specific type of treatment is effective or not. However, in practice, most diseases are treated with more than one drug. For example, hypertension is usually managed by a combination of different classes of drugs such as angiotensin receptor blockers/angiotensin-converting enzyme inhibitors, beta-blockers, calcium channel blockers, and diuretics. Policy-makers, physicians, and patients need to be able to select the best treatment among the many potential options. Direct comparisons between two or more therapies are not always possible, but indirect data is usually accessible. Different therapies may have been

A. Kumar, *Meta-analysis in Clinical Research: Principles and Procedures*, https://doi.org/10.1007/978-981-99-2370-0_11

Fig. 11.1 Simple Network

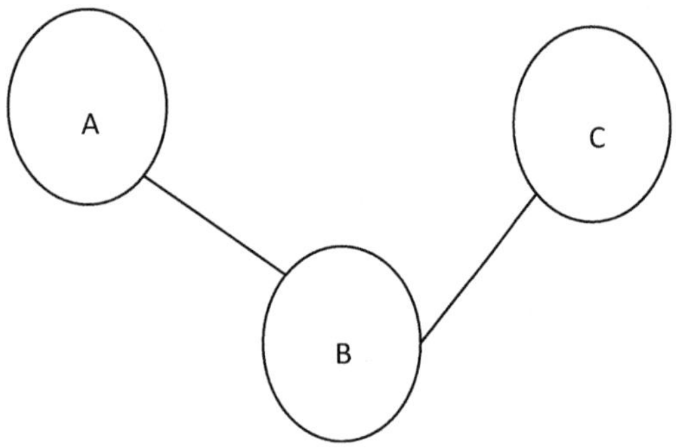

tested in different trials, but the control group may have been the same in all of them. If two treatments were never directly compared, but the efficacy of both medications compared to placebo has been well researched, it is possible to compare both medications through a network meta-analysis. A network meta-analysis compares several treatments, utilising direct comparisons of interventions within randomised controlled trials as well as indirect comparisons across trials based on a common comparator. Mixed-treatment comparison meta-analysis is another name for network meta-analysis (White 2015; Lumley 2002).

This is due to the fact that it combines several direct and indirect treatment comparisons into a single model, which can be described as a 'network' of comparisons. A network of interventions is made up of research that links three or more interventions through direct comparisons. Multiple techniques to perform indirect comparisons between interventions can be found in a network of interventions. These are indirect comparisons that can be generated using mathematical combinations of the available direct intervention impact estimates.

In the most basic form of a network meta-analysis, two treatments of interest have been directly compared to a common comparator, but not to one another. Typically, placebo or standard therapy will be used as a comparator. Figure 11.1 illustrates this condition. The circles represent therapies, while the lines represent treatment comparisons in head-to-head clinical trials. Treatments A and C have been directly compared to treatment B in this example, but no trials have been conducted to compare treatments A and C. This chapter delves into network meta-analysis in depth.

11.2 Network

To begin, we must define what we mean by a 'network' of therapies. Assume we have gathered data from a randomised controlled experiment that compared the effect of therapy A to that of another condition B, as shown in Fig. 11.1.

If you look closely at Fig. 11.1, there are two core components. The first one is two circles (the so-called nodes), which represent the two conditions A and B in the trial. The second component is the line connecting these two nodes. This line is called an edge. The edge represents how A and B are related to each other. The relationship between A and B can be described in suitable effect sizes such as odd ratio, mean difference, etc. depending upon the outcome measure. Suppose we also collected the data from another study that used the control condition of B but instead of the administration of A drug, this study used another drug C. In this particular study, treatment C was also compared with treatment B. Therefore, it is clearly visible from Fig. 11.1, two effect sizes comparing A to B, and C to B. This information we call direct evidence as these effect sizes were directly observed in real trials. Suppose B was used as a control group.

In Fig. 11.1, all nodes are either directly or indirectly connected. The B treatment (control group) is directly connected to all the other nodes. It takes only one 'step' in the figure to get from B to the two other nodes A and C. B → A, B → C. In contrast, A and C only have one direct connection, and they both connect to B. A → B and C → B. However, there is an indirect connection between A and C. This connection exists because B serves as the link, or bridge, between the two conditions: A → B → C. As a result, there is indirect evidence for the relationship between A and C, which can be derived from the structure of the network. Using information from the directly observed edges, we can calculate the effect of the indirectly observed comparison between A and C. A direct comparison is referred to as a closed loop. When there is no closed loop, direct comparison is not possible, and indirect comparison is required.

11.3 Network Meta-Analysis

Network meta-analysis, also known as multiple treatment meta-analysis or mixed-treatment comparisons, is a popular method for combining information from several studies (typically randomised trials) that compare different therapies (or other interventions). Its most distinguishing feature is its capacity to incorporate direct and indirect data; for example, when comparing treatments, A and B, studies that directly compare A with B (direct evidence) as well as studies that compare A with C and B with C are used (indirect evidence). A network meta-analysis can contain an unlimited number of treatments, trials, and patients. To conduct a typical network meta-analysis, however, the treatments must form a connected network, with a path connecting each therapy to every other treatment (Jansen et al. 2011).

11.4 Direct and Indirect Evidence in a Treatment Network

Direct evidence is information derived from comparisons that have been studied in the research that has been included. By reducing the effect of one (directly observed) comparison from the effect of a related comparison, indirect evidence can be inferred from direct evidence (e.g. a comparison that used the same control group) (Bucher et al. 1997; Lu and Ades 2004).

11.5 Benefits

In network meta-analysis, direct and indirect evidence are combined in one model. We can estimate the (relative) effect of each included treatment using this information. Even when there is direct evidence for that specific comparison, we can improve the precision of an impact size estimate by adding indirect evidence. Overall, network meta-analysis has various advantages, as listed below (Salanti 2012).

- It enables us to combine all the available data from a group of similar studies into a single analysis. Trials evaluating different treatments in a typical meta-analysis must pool each comparison (e.g. treatment A vs. placebo, treatment B vs. placebo, treatment A vs. treatment B, etc.) in a distinct meta-analysis.
- In contrast to traditional meta-analysis, network meta-analysis allows for the inclusion of indirect evidence in a network. Only direct evidence from comparisons that were actually included in a trial can be pooled in pairwise meta-analyses.
- Network meta-analyses allow us to infer which sort of treatment is superior for the target population under research if all assumptions are met and the results are sufficiently conclusive.
- A network meta-analysis makes use of all direct and indirect evidence available. Compared to a single direct or indirect estimate, empirical investigations imply that it offers more precise estimates of intervention impacts.
- Furthermore, network meta-analysis can give data for comparisons between pairings of therapies that have never been studied in separate randomised trials. The evaluation of their relative ranking for a given outcome is made possible by comparing all relevant interventions in the same analysis.
- Conditions with multiple interventions.
- Many combinations of direct or indirect interactions.
- To answer more relevant clinical questions.
- To make treatment estimates for an entire treatment network instead of scanning each individual pairwise comparison.
- To give the 'full picture' to clinicians, Gain precision by considering all available evidence, not just (A vs. B comparisons),
- Potential to more explicitly 'rank' treatments using summary outputs.

11.6 Network Meta-Analysis Models

As an example, we used a simple network with three nodes and edges. In practice, however, a network meta-analysis frequently includes a substantially larger number of treatments. This quickly leads to much more complicated networks. As a result, we require a computational model that allows us to efficiently and internally consistent pool all accessible network data. For network meta-analysis, several statistical techniques have been developed.

While network meta-analysis models may differ in their statistical approaches, the good news is that when the sample size is large enough, they should all generate the same results. In general, no network meta-analysis method is better or worse than the others. As a result, you can securely choose one of the two approaches. Frequentist inference methods are (still) far more common in most disciplines than Bayesian approaches. As a result, certain people may find it easier to comprehend the types of outcomes given by a frequentist model.

In practice, it is a good idea to utilise one method for the main analysis and the other method for the sensitivity analysis. If the two methods arrive to the same conclusion, we can be more confident that the results are reliable.

A frequentist or Bayesian method of network meta-analysis is conceivable. In practice, each of these strategies has its own set of advantages, but the overall outcomes are often relatively similar. We can also include study covariates that predict effect size differences in network meta-analyses based on a Bayesian hierarchical model. As a result, a network meta-regression model is created. In our network, indices like the SUCRA or *P*-score can be utilised to see which sort of treatment is the most beneficial. It is, nonetheless, critical to include uncertainty into our decision-making process. Different treatments' confidence/credible intervals frequently overlap, making it difficult to determine whether one format is actually superior than all others. A frequentist or Bayesian model can be used to undertake network meta-analysis. Both models are equal and provide similar results as the sample size grows (Madden et al. 2016; White 2015).

11.7 Network Meta-Regression

Network meta-regression is useful to see if certain research variables have an impact on the magnitude of effect sizes in our network. It can also be used to look for variables that could be causing the inconsistency. Assume that we want to see if a study's risk of bias has an impact on the effects in our network meta-analysis. We can adjust for such a relationship and examine its impact on our results by integrating the risk of bias as a predictor in our model. One of the most important questions in network meta-analysis is whether the network is consistent—that is, whether the direct and indirect evidence agree (and, if there are multiple sources of indirect evidence, whether they agree with each other). Inconsistency statistical models have been proposed that can be used to analyse consistency (Donegan et al. 2018).

11.8 Assumptions

A good network meta-analysis assumes that the diverse sets of studies included in the analysis are similar, on average, in all key factors that may influence the relative effects. When several sources of information (e.g. direct and indirect) about a particular intervention comparison disagree, this is known as incoherence (also known as inconsistency). The first step in grading the evidence from a network meta-analysis is to assess confidence in each direct comparison.

The overall confidence in the evidence is determined by combining domain-specific ratings. Network meta-analysis is a valuable tool for estimating the relative effectiveness of several therapies or interventions in a group setting. Network meta-analysis uses both direct (observed) and indirect evidence to estimate treatment effects. This, however, is predicated on the transitivity assumption. When we can integrate direct evidence from two comparisons to get valid indirect evidence about a third, we have transitivity. Consistency is the statistical representation of transitivity, while inconsistency is the polar opposite. When the genuine effect of a comparison based on direct evidence differs from the one based on indirect evidence, inconsistency occurs.

Inconsistencies in our network can be identified using techniques like node splitting and net heat maps. When inconsistencies are discovered, the validity of our entire set of results is jeopardised. In such circumstances, the entire network should be examined for traits that may have contributed to systematic disparities in study/design results.

For network meta-analysis to be valid, three assumptions must be met.

1. Homogeneity of direct evidence
2. Transitivity, similarity, or exchangeability
3. Consistency between direct and indirect evidence

11.8.1 Homogeneity of Direct Evidence

In any meta-analysis, including network meta-analysis, the homogeneity of treatment effects across trials reduces confidence in pooled results. Direct evidence (within each pairwise comparison) in network meta-analysis should be sufficiently homogenous. Standard approaches for evaluating homogeneity can be used to analyse the homogeneity of direct evidence are Forest plots, I^2, and the Cochran Q test.

11.8.2 Transitivity

Participants in networked trials of various comparisons must be sufficiently similar. This is characterised in terms of effect modifier dispersion. Patients should be similar enough that they have an equal chance of receiving any of the therapies in the network. To attribute treatment results to intervention, active and placebo-controlled interventions must be sufficiently similar across trials. Although transitivity cannot be practically tested, it should be understood philosophically. The transitivity assumption states that direct evidence can be used to infer unobserved, indirect evidence and that direct and indirect evidence are consistent.

11.8.3 Consistency (Indirect and Direct)

For the same treatment comparison, consistency refers to the agreement between direct and indirect estimates. To achieve consistency in closed loops, direct and indirect estimates of treatment effects should be identical (previously known as coherence). Several reasons might cause inconsistency, including variances in patients, treatments, settings, and timing. Consistency models are statistical models that presume consistency and allow for direct and indirect estimate discrepancy (inconsistency models). Consistency may and should be measured objectively. The presence of significant inconsistency suggests that network meta-analysis should be avoided.

When there is no closed loop in the evidence network, simple indirect comparisons are used. The point estimates and overlap of confidence intervals are qualitatively compared in the simple indirect comparison approach. If the treatments' direct effects on a common comparator (e.g. placebo) are similar in direction and amplitude, with high confidence interval overlap, they are likely to be equivalent in effectiveness. Because confidence interval overlap is not a valid substitute for rigorous hypothesis testing, such results must be treated with caution.

The ability to do network meta-analyses requires transitivity, which cannot be assessed directly. Consistency is a statistical form of transitivity that occurs when impact size estimates based on direct evidence are the same or similar to estimates based on indirect evidence (Donegan et al. 2013).

11.9 Limitations

Every method has some limitations. Some of the limitations associated with network meta-analysis are mentioned below (Cipriani et al. 2013):

- Requires sophisticated statistical knowledge and software.
- Assumes that all of the interventions in the 'network' are equally applicable to all populations and study circumstances.
- There is a chance that study selection bias will be introduced.
- The ease with which researchers can conduct this type of review may motivate them to do so even if it is not the best study design for their research issue.
- Methodological techniques that are complicated.

11.10 Conclusion

A network meta-analysis can be used to compare two interventions that have not been studied together in the same trial. Network meta-analysis is a method for assessing three or more therapies in one study using both direct and indirect data from a network of research. Network meta-analysis generates estimates of the relative effects of each pair of treatments in the network, which are frequently more exact than a single direct or indirect estimate. It also enables the assessment of intervention ranking and hierarchy.

The systematic review must be prepared thoroughly and done properly to ensure the validity of findings from network meta-analyses. Defining the review question, specifying eligibility criteria, searching for and selecting studies, assessing the risk of bias and quality of evidence, conducting a network meta-analysis, and interpreting and reporting findings are all aspects of designing and conducting a systematic review for network meta-analysis.

References

Bucher HC, Guyatt GH, Griffith LE et al (1997) The results of direct and indirect treatment comparisons in meta-analysis of randomized controlled trials. J Clin Epidemiol 50:683–691

Cipriani A, Higgins JP, Geddes JR et al (2013) Conceptual and technical challenges in network meta-analysis. Ann Intern Med 159(2):130–137

Donegan S, Williamson P, D'Alessandro U et al (2013) Assessing key assumptions of network meta-analysis: a review of methods. Res Synth Methods 4(4):291–323

Donegan S, Dias S, Tudur-Smith C et al (2018) Graphs of study contributions and covariate distributions for network meta-regression. Res Synth Methods 9(2):243–260

Jansen JP, Fleurence R, Devine B et al (2011) Interpreting indirect treatment comparisons and network meta-analysis for health-care decision making: report of the ISPOR task force on indirect treatment comparisons good research practices: part 1. Value Health 14:417–428

Lu G, Ades AE (2004) Combination of direct and indirect evidence in mixed treatment comparisons. Stat Med 23:3105–3124

Lumley T (2002) Network meta-analysis for indirect treatment comparisons. Stat Med 21(16): 2313–2324

Madden LV, Piepho HP, Paul PA (2016) Statistical models and methods for network meta-analysis. Phytopathology 106(8):792–806

Salanti G (2012) Indirect and mixed-treatment comparison, network, or multiple-treatments meta-analysis: many names, many benefits, many concerns for the next generation evidence synthesis tool. Res Synth Methods 3:80–97

White IR (2015) Network meta-analysis. Stata J 15(4):951–985

Registration and Software

12

Abstract

Prospective registration of protocols of systematic reviews and meta-analysis with international organisations like PROSPERO is always recommended to reduce duplication and provide more clarity regarding aim and objectives as well as analysis of data and bias. Various user-friendly software are available freely as well as commercially for the analysis. Every software has its own advantages and limitations. This chapter provides a detailed discussion regarding the registration of protocols for meta-analysis as well as commonly used software in meta-analysis.

Keywords

Registration · PROSPERO · RevMan · STATA

12.1 Introduction

Registration of the protocol of the meta-analysis is an important step to ensure the quality of the meta-analysis. Various organisations are available where researchers can register their protocols. One of the well-known organisations is PROSPERO. Meta-analysis can be performed using a variety of free and commercial software. In this chapter, I have discussed in detail regarding registration and software used in meta-analysis.

A. Kumar, *Meta-analysis in Clinical Research: Principles and Procedures*, https://doi.org/10.1007/978-981-99-2370-0_12

12.2 Why Is a Registry Needed?

There are a huge number of systematic reviews published in the literature. However, all of them are not registered. Excessive duplication of systematic reviews is common and will continue to occur in the absence of a registry. It has been found that on the same topic many systematic reviews have been published. The reviews differed in their recommendations and reporting quality. Nonetheless, I wonder if there is a need for so many systematic evaluations of a single topic. Such duplication of effort comes at a huge expense in terms of human and financial resources that may be better spent on other research or health-care delivery. Unpublished reviews, like any unpublished research, may include potentially useful information.

Many institutions and organisations are debating how to prioritise systematic review completion, aiming to strike a balance between the requirements of patients, doctors, and other decision-makers and the resources available. Given that many organisations require reviews on the same issues (such as drug and technology evaluations), minimising unnecessary duplication would allow these agencies to focus their efforts on other areas of research. A register of reviews would make it easier for decision-makers and scholars to see which reviews are currently being worked on by other organisations. We hope that this information will facilitate collaboration by bringing together research groups with similar interests and, potentially, by increasing the number of team members engaging in the review process.

Because the duty of updating might be shared among groups of academics, improved collaboration could also make it easier to update reviews. The complexity of updating systematic reviews, as well as the need for harmonising various components of the updating process, is recognised by producers and financiers of systematic reviews. Rapid evaluations have grown in popularity in recent years as a result of the requirement for quick decisions in political and other spheres. A database of existing evaluations is likely to be a good place to start when making such a hasty decision. Finally, registering systematic review processes would aid reviewers and editors in detecting bias in outcome reporting.

The goals of a registry of systematic reviews would be to reduce duplication, provide more clarity surrounding the review process and analysis, eliminate publication bias and selective reporting of outcome-related bias, foster collaboration, and assist with prioritisation. We suggest that the registry's initial focus be on reviews of health-care interventions, as these are the most often reported and have the most advanced methodologies. The registry should be expanded in the future to include all systematic reviews. We feel that this register should be included in existing clinical trial registries' plans for utilising existing technology and resources. A register of trials and reviews that coexists could encourage trialists and reviewers interested in the same issues to collaborate.

For each review, a minimum data set should be collected. Upon completion of registration, the investigator will be given a registration number. An automated warning would be issued to the investigator every year, requesting an update, allowing the site to remain maintained. Completed review citations will be included when they become available from the writers. End-users would benefit from open access if entire reviews were available. The registration should be completed by the investigators who initiated the study. The database's contents should be freely available and searchable.

Editors of journals might encourage systematic review registration and demand a registration number before considering an article, similar to how clinical trials are registered. This would be a good safeguard for journal editors in their ongoing attempts to improve transparency by preventing selective reporting of results. Financing agencies might also make registration a requirement of their funding processes, furthering their open access and transparency goals.

Systematic review registration will require more work from reviewers to submit their protocols, will use more resources for expanding the present procedure of registering clinical trials, and will not alleviate all of the challenges that exist in the publication of health-care research. However, if it is pushed, endorsed, and followed, it is likely to reduce publication bias and boost transparency (Xu et al. 2019; Tawfik et al. 2020).

12.3 When to Register?

The registration number in a meta-analysis refers to the registration of the meta-analysis protocol. Before beginning the meta-analysis, and specifically before beginning the data extraction, you should usually register your protocol. This is for two reasons.

- When registering your protocol, you can see if any similar studies are being carried out so that you do not repeat them. (Keep in mind that your meta-analysis will address a very specific question, and you may opt to look at data and datasets that other researchers may not have.) You could potentially get to various conclusions based on the same evidence.
- This also eliminates the risk of data-driven skewing of the results.

You can still try to register your protocol, though. If you still have time until your study is over, you may be able to get your registration number before then and submit it with your final paper to the journal.

12.4 Some Registry Sites

PROSPERO: An open registry for all types of systematic reviews (The International Prospective Register of Systematic Reviews (PROSPERO) 2022).

Cochrane: A registry for systematic reviews of health-care interventions (The Cochrane Database of Systematic Reviews (CDSR) 2022).

Campbell Collaboration: A registry for systematic reviews of social interventions (Campbell Collaboration 2022).

INPLASY: The International Database to Register your Systematic Reviews (International Platform of Registered Systematic Review 2022).

12.5 Steps Involved

1. Creation of account in the portal.
2. Addition of all relevant information related to study.
3. Submission of relevant information.
4. Reviews by the organisation.
5. Resolving of queries if any.
6. Assignment of registration number.

12.6 Software

A suitable software is required to conduct a meta-analysis. Broadly, three types (spreadsheet, statistical package, and stand-alone packages) are being used. The most basic calculations of meta-analysis such as calculations of weighted averages can be done in a spreadsheet by a knowledgeable user. The spreadsheet is most commonly used for calculations of meta-analysis; however, it is prone to transcription errors and many advanced calculations cannot be done which is required nowadays. Therefore, spreadsheets are no longer used nowadays for meta-analysis purposes. Various statistical packages are available nowadays for various purposes. Some of these packages are also useful to perform meta-analyses such as SAS, SPSS, Stata, S-Plus, R, and BUGS. The complex analysis such as multivariate, Bayesian, etc. can also be done using these packages; however, users should have the technical knowledge to use these programmes in a better way.

Nowadays, stand-alone packages specifically for meta-analysis are more popular among researchers. The well-known and most commonly used stand-alone packages are Comprehensive Meta-Analysis (CMA), MetaWin, MIX, MetaAnalyst, and RevMan. These packages are easy to use to perform all types of calculations related to meta-analysis. The procedures of these software vary among each other's in terms of importing and exporting of data, effect sizes calculations, statistical models, subgroup analysis, sensitivity analysis, and so on. For example, we are unable to do a quantitative analysis of publication bias using RevMan. These packages also differ in user-friendliness, technical support, documentation, and price.

Various user-friendly software are available for meta-analysis such as RevMan, STATA, R, etc. I have discussed in detail about most commonly used software in a meta-analysis that is RevMan.

12.6.1 RevMan

RevMan 5 is free of charge for preparing Cochrane and non-Cochrane Reviews (for purely academic use). It is available for download as locally installed software (current version: 5.4.1), released in September 2020. The only thing is researchers using RevMan for their analysis should be properly cited (The Cochrane Collaboration 2020). There are two versions of RevMan. The first one is RevMan web which is specifically for Cochrane systematic reviews and is commercially available, whereas the other version is to download and is most commonly used for academic purposes.

12.6.1.1 Steps

1. The First step is to download RevMan 5 which is available freely at https://training.cochrane.org/online-learning/core-software-cochranereviews/revman/. However, before downloading, important information is must to provide by the user like name, job role, Academic affiliation, Email address, Country, as well as a short description of the work you expect to do with RevMan. The snapshot is mentioned in Fig. 12.1.
2. After submitting the information, RevMan will ask which mode you would like to use? There are two modes, i.e. Cochrane and non-Cochrane mode. For Cochrane systematic review, the server will automatically connect to Cochrane, whereas in-non-Cochrane systematic review, the server will not be connected to Cochrane. The next step is to install the RevMan as per the configuration of the system as mentioned in Fig. 12.2.
3. Next step is to open installed RevMan 5.4. After opening, the window will show 'welcome to Review Manager 5.4' as shown in Fig. 12.3.

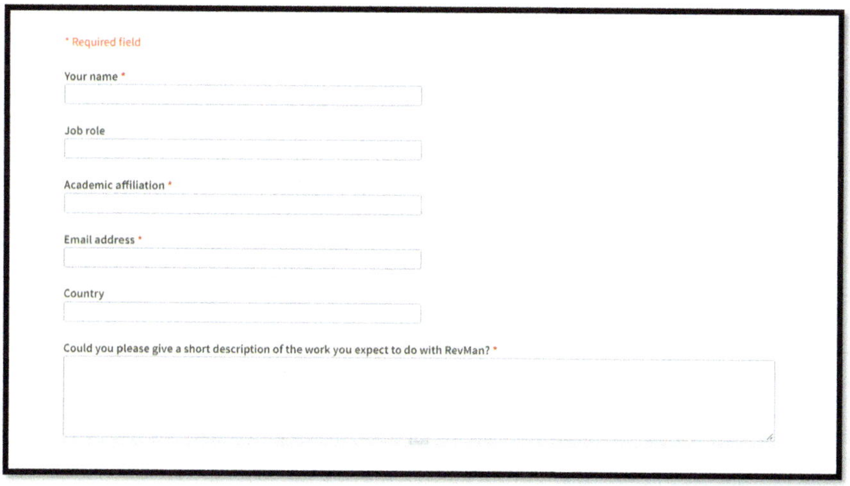

Fig. 12.1 Snapshot of the required information to fill up before downloading of RevMan

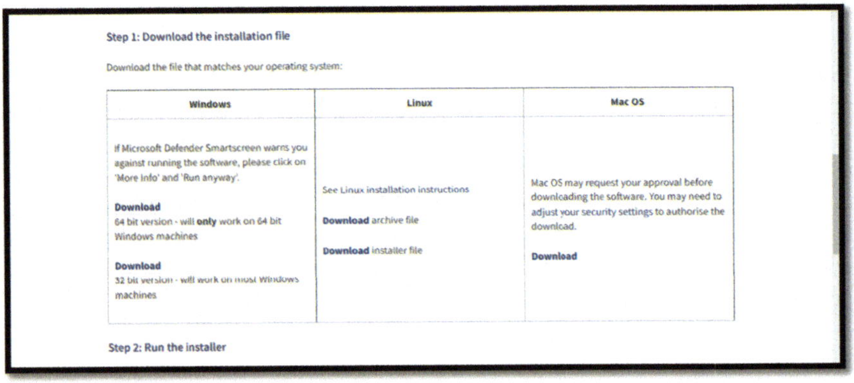

Fig. 12.2 Snapshot of downloading RevMan 5 as per the configuration of the system

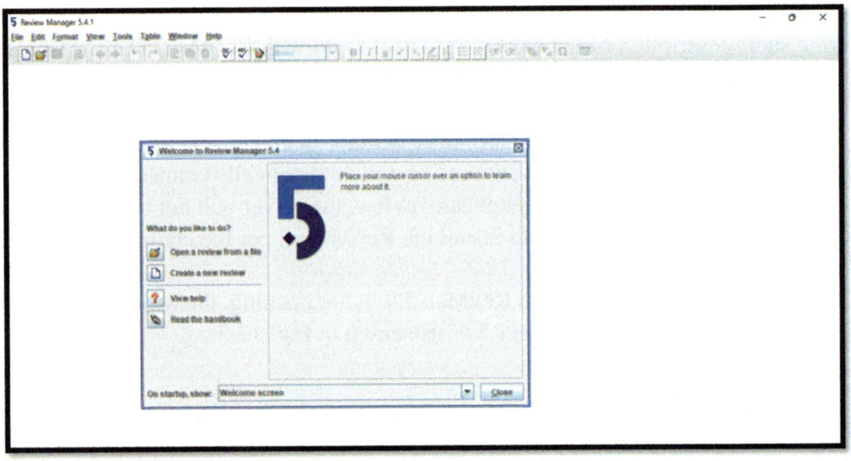

Fig. 12.3 Snapshot of RevMan 5 (after opening)

4. Next step is to click on create a new review. After that, you can see a tab showing 'Welcome to the New Review Wizard' as shown in Fig. 12.4.
5. You can see a New Review Wizard after clicking on the next. The New Review Wizard will ask you Which type of review do you want to create? You will see the following options: Intervention review, Diagnostic test accuracy review, Methodology review, Overview of reviews, and flexible reviews. Researchers could select the most appropriate option as per the aim and objectives of the review. For example, as continued with our previous example regarding the use of steroids or use of aspirin in the reduction of mortality of COVID-19 patients. This is a classic example of an Intervention review. Therefore, I have selected an 'intervention review' and clicked on next as shown in Fig. 12.5.

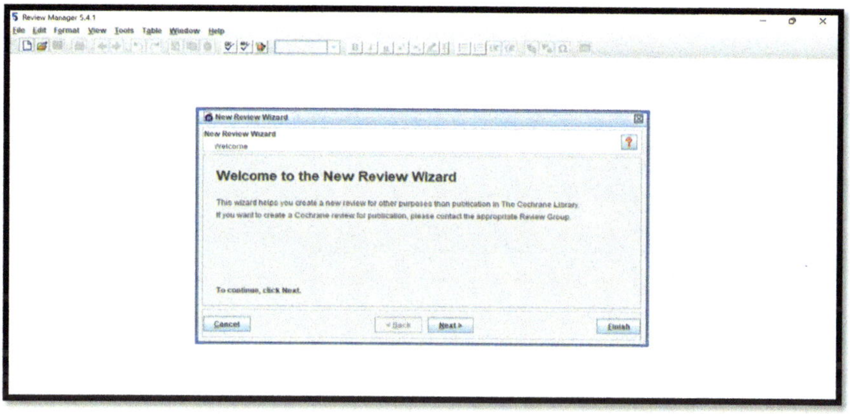

Fig. 12.4 Snapshot of RevMan 5 (after clicking on new review)

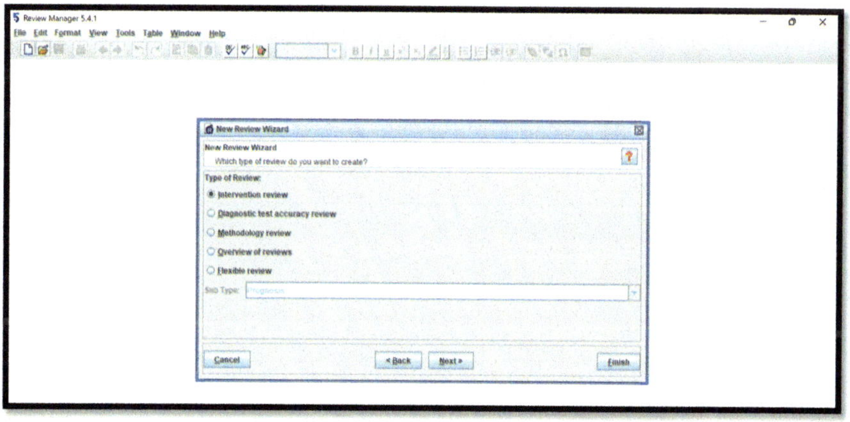

Fig. 12.5 Snapshot of New Review Wizard

6. Click on next and a new box will appear where the title of the review will be added by the researchers as per the objectives of the review. I have selected the second option as per the aim of our study where we would like to compare the role of steroids as compared to non-steroids in the reduction of mortality of COVID-19 patients as shown in Fig. 12.6.
7. Click on next, it will ask which stage (protocol or full review) should the review start in? (Fig. 12.7). I have selected full review and click on finish.
8. After clicking on the finish, you will see the RevMan as shown in Fig. 12.8. The next step is to write the protocol of the review. You can see on the left-hand side where many options such as Authors, Contact persons, Background, Description of the condition, Description of the intervention, How the intervention might work, Why it is important to do this review, Objectives, Methods,

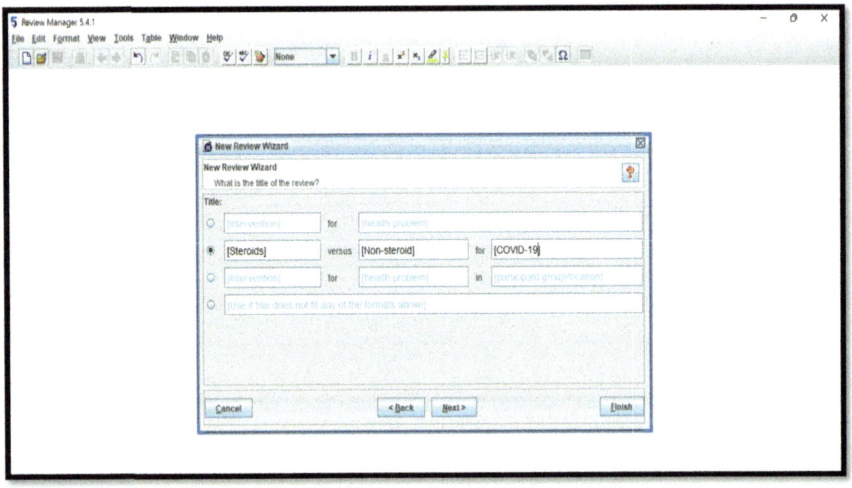

Fig. 12.6 Snapshot of addition of title of review as per the objectives

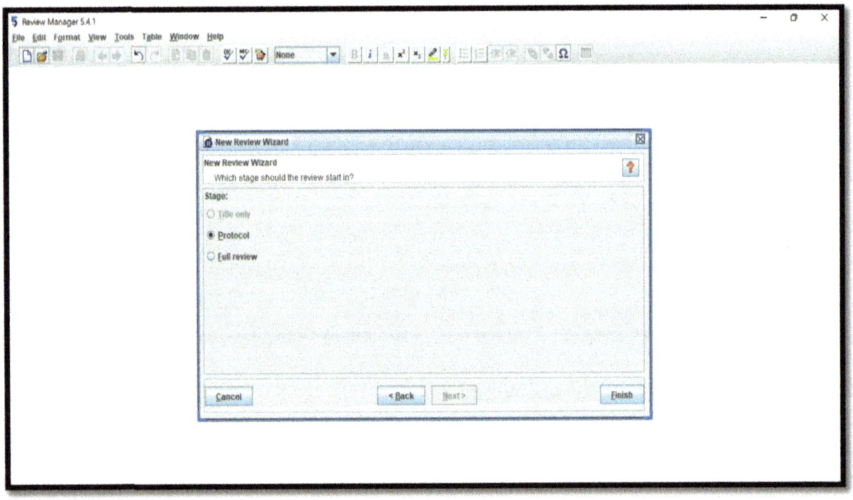

Fig. 12.7 Snapshot of selection of stage of review

Criteria for considering studies for this review, Types of studies, Types of participants, Types of interventions, Types of outcome measures, Primary outcomes, Secondary outcomes, Search methods for identification of studies, Electronic searches, Searching other resources, Data collection, and analysis, Selection of studies, Data extraction, and management, Assessment of risk of bias in included studies, Measures of treatment effect, Unit of analysis issues, Dealing with missing data, Assessment of heterogeneity, Assessment of

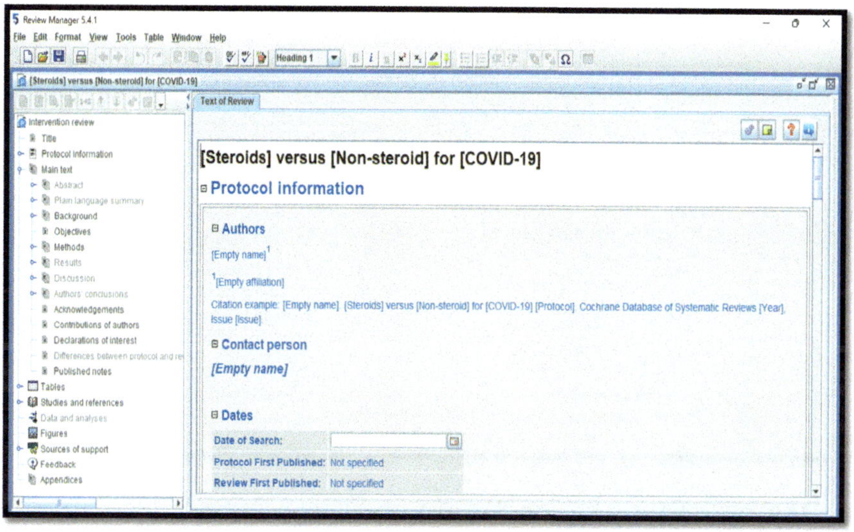

Fig. 12.8 Snapshot of RevMan after selection of stage of review as a full review

reporting biases, Data synthesis, Subgroup analysis and investigation of heterogeneity, Sensitivity analysis, Acknowledgement, Contributions of authors, Declarations of interest appear. If you see carefully, you will find some of the options are disabled particularly results, abstract, plain language summary, discussion, and authors conclusion.

9. Next step is to import studies into the RevMan. There are a couple of ways for importing of studies. You can import references as PubMed files of the article. Click on the file and select import followed by references as shown in Fig. 12.9.
10. Next tab is opened as 'Import references Wizard' as shown in Fig. 12.10.
11. After clicking on the next, you can select the reference file of the article downloaded from PubMed as shown in Fig. 12.11.
12. After clicking on open, the next box will appear as 'Import References Wizard' as shown in Fig. 12.12, and click on the next.
13. After clicking on next, a new box will appear as 'Where should the references be stored' with the following options: Included studies, excluded studies, Studies awaiting classifications, ongoing studies, additional references, other published versions of this review, classification pending references. In continuation of our previous example, we have selected the option 'included studies' as shown in Fig. 12.13.
14. After clicking on next, a new box will appear which will ask you Do you want to exclude any references or to define the destination for each individually? as shown in Fig. 12.14. If not, click on finish.
15. After clicking on finish, one new message appears as '1 reference imported' as shown in Fig. 12.15.

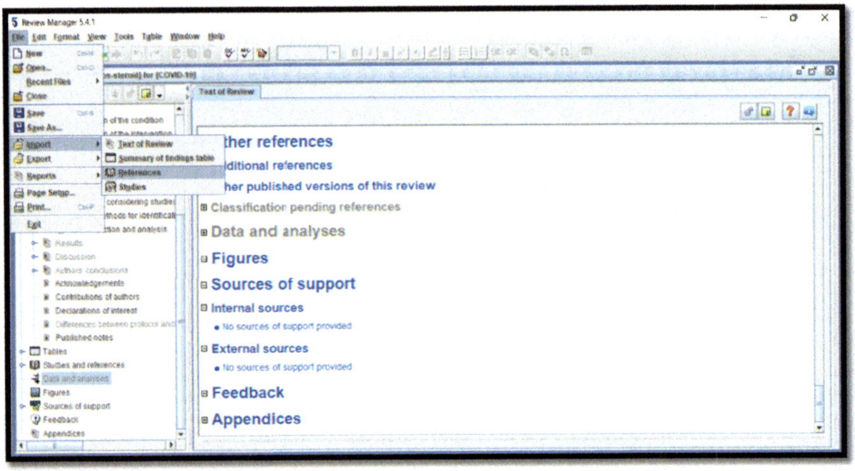

Fig. 12.9 Screenshot of import of studies into the RevMan

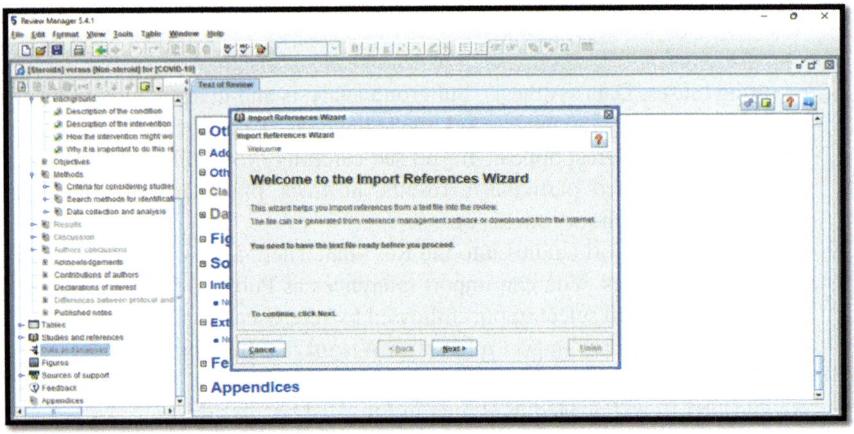

Fig. 12.10 Screenshot of Import References Wizard

16. Next step is data analysis. Just to make the reader better understand, I have added two studies in the RevMan so that I can show you the analysis. A new wizard will open as 'New Comparison Wizard'. The question will appear as 'What name should the comparison have?' as shown in Fig. 12.16.
17. Add the name of the comparison. In our example, we would like to compare the effect of steroids with the non-steroidal group in the reduction of mortality of COVID-19 patients. Therefore, in the comparison box, we have added 'steroid vs non-steroid' as shown in Fig. 12.17.

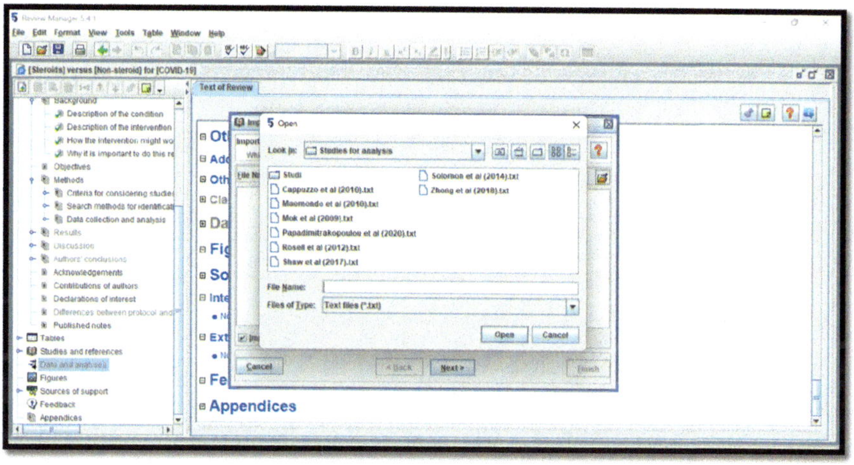

Fig. 12.11 Screenshot of Import of references into the RevMan

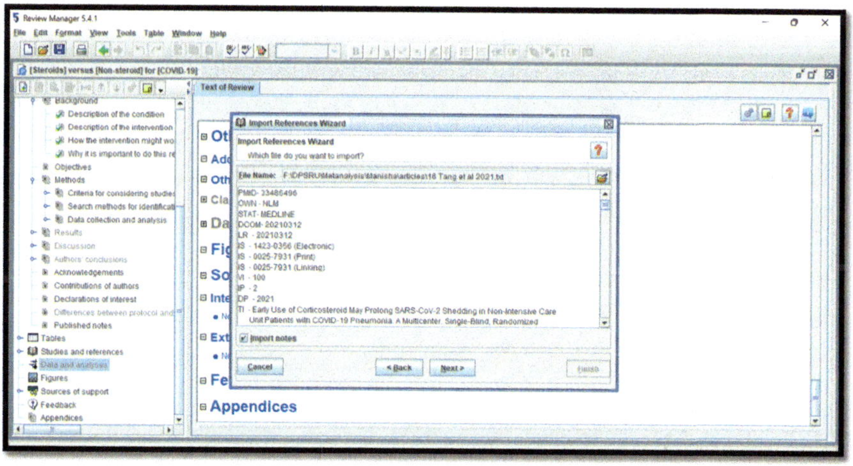

Fig. 12.12 Screenshot of Import References Wizard

18. After clicking on next, a new question will appear as 'What do you want to do after the wizard is closed' In our example, we would like to add an outcome under the new comparison, therefore, selected as shown in Fig. 12.18.
19. New Wizard will open as 'New Outcome Wizard' after clicking on the continue option. The question will be appeared as 'What type of outcome do you want to create' The following options will appear under the type of data: Dichotomous, Continuous, O-E and Variance, and other data. The researcher should select the appropriate option as per the type of data extracted from individual studies. In our example, as we would like to find out the role of steroids in the reduction of

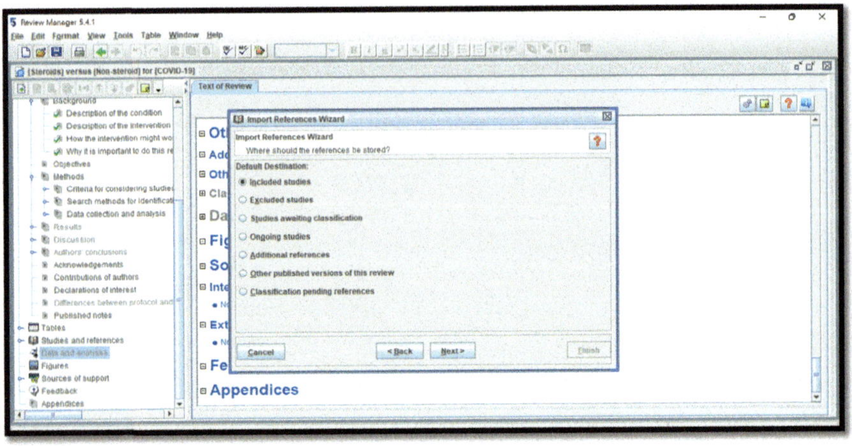

Fig. 12.13 Screenshot of Import References Wizard (selection of included studies)

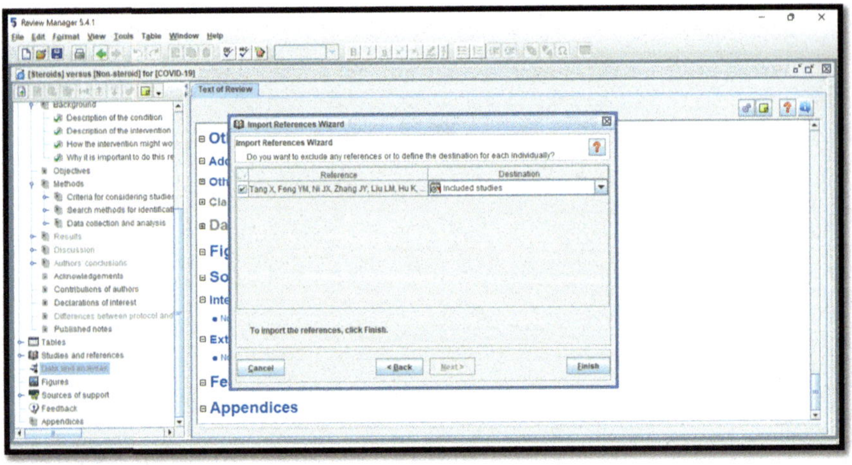

Fig. 12.14 Screenshot of import references wizard (last step)

mortality of COVID-19 patients, therefore data is dichotomous in nature and we have selected the option "Dichotomous" as shown in Fig. 12.19.

20. The next wizard will appear as "What name should the outcome have?" with the following options: Name, Group label 1, and group label 2. In our example, I have added the outcome as "death" as we would like to know the effect of steroids on the deaths of COVID-19 patients. The labels are also mentioned as steroids and non-steroids as shown in Fig. 12.20.
21. Next wizard will ask you "Which analysis method do you want to use?". You have to choose a suitable statistical method, analysis model, and effect measure depending on the objectives of your study. In our case, we have selected the

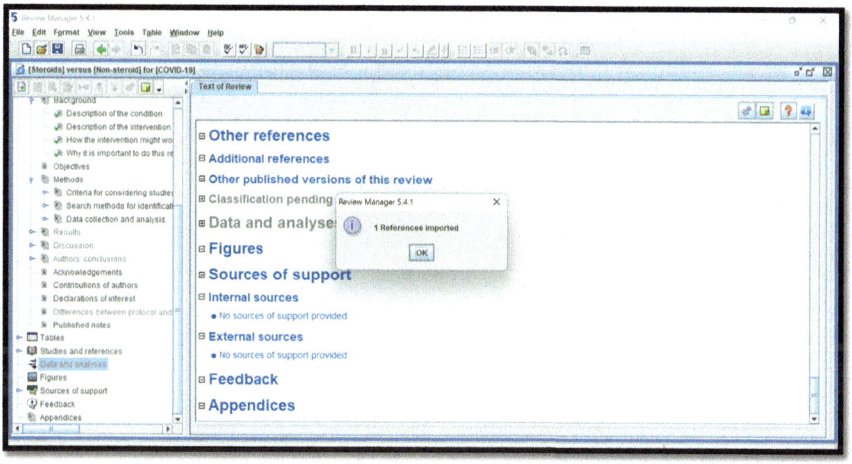

Fig. 12.15 A screenshot of the message appears after importing studies in the RevMan

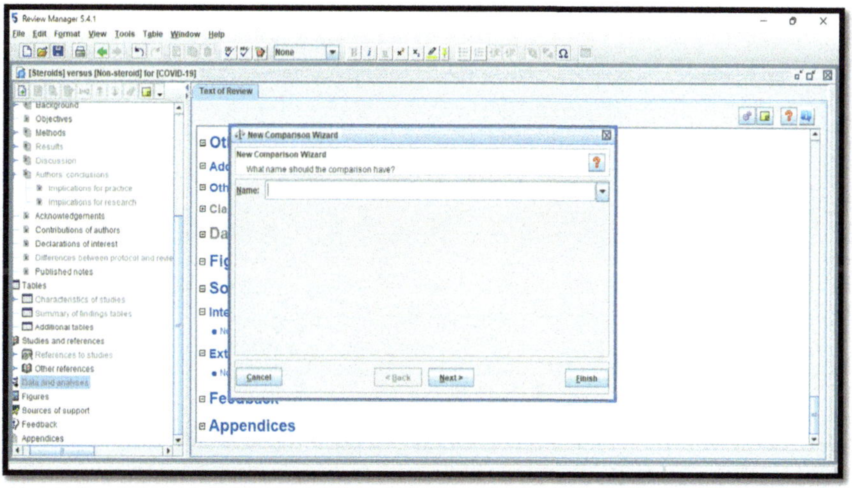

Fig. 12.16 Screenshot of new comparison wizard

statistical method as "Mantel-Haenszel", analytical model as "fixed effect" and effect measure as "odd ratio" as shown in Fig. 12.21. The detail of the analysis of data is presented in Chap. 4.

22. After clicking on the next, a new wizard will be opened with the question "Which analysis details do you want to use?" Usually, we are selecting a 95% confidence interval as shown in Fig. 12.22, however, it will depend upon your aim and objective of the study.
23. After clicking on next, a new outcome wizard will appear. You can add a subgroup for the new outcome, or can add study data for the new outcome or

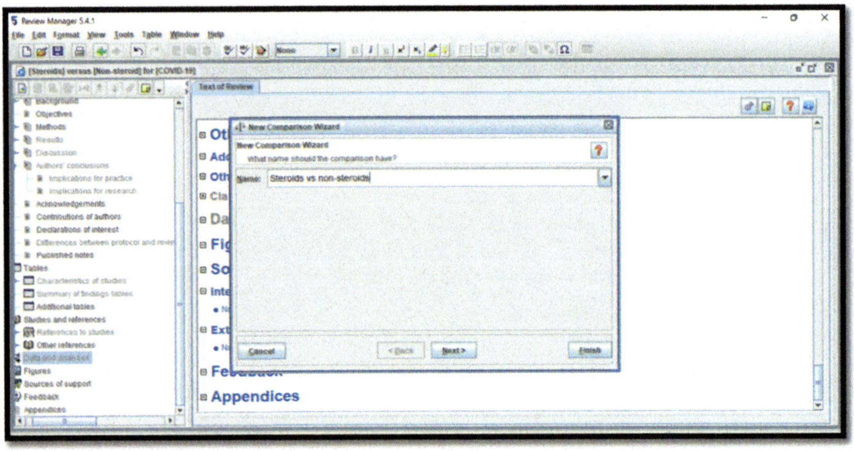

Fig. 12.17 Screenshot of new comparison wizard after adding suitable example

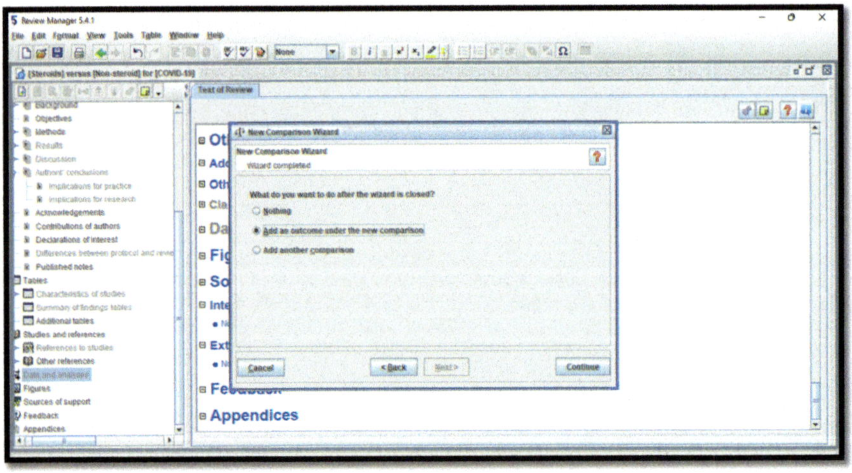

Fig. 12.18 Screenshot of New Comparison Wizard to add outcome under the new comparison

add another outcome for the same comparison, or can finish depending upon the aim and objectives of your study as shown in Fig. 12.23. In our example, we do not want a further outcome, so clicked on the finish.

24. Next step is to open new study data wizard and selection of desired studies as shown in Fig. 12.24.
25. The table will appear to add data from individual studies manually. In our example, the column of tables appears as "events and total" in both steroidal and non-steroidal groups" as shown in Fig. 12.25.

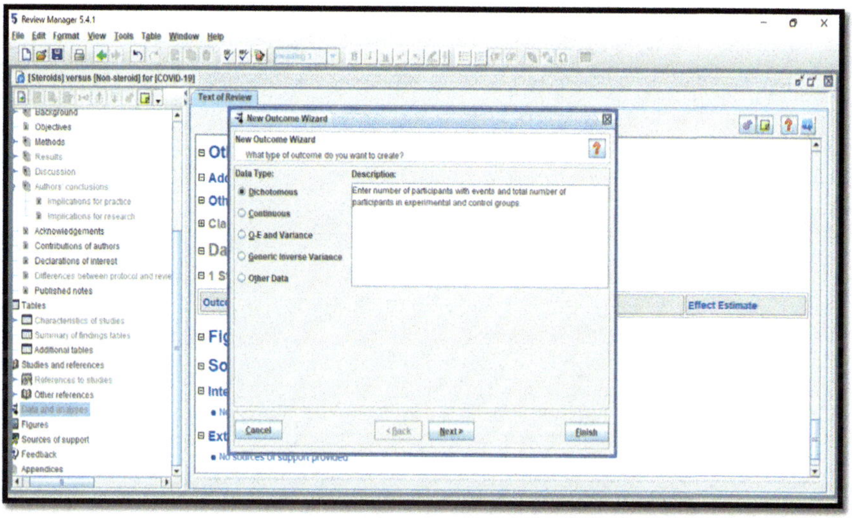

Fig. 12.19 Screenshot of selection of data

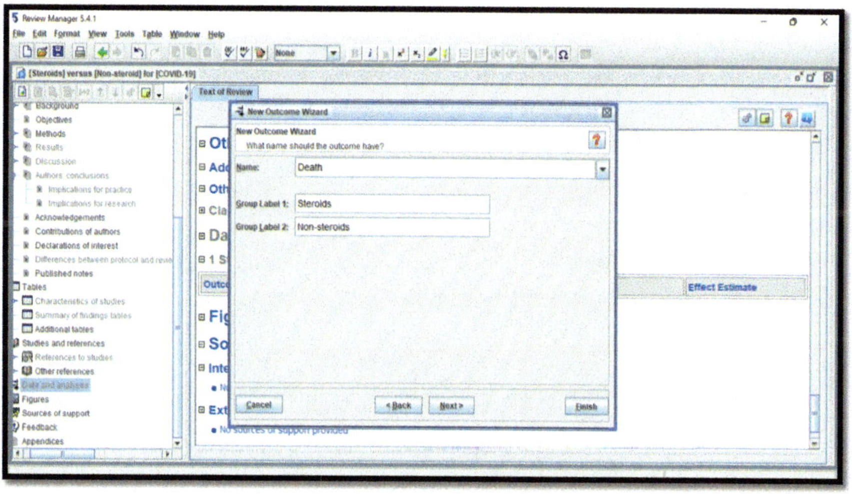

Fig. 12.20 Screenshot of New Outcome Wizard to add the name of outcome and group labels

26. I have added hypothetical numbers for a better understanding of readers. You can see after the addition of numbers, automatically forest plot is created on the right-hand side as shown in Fig. 12.26.
27. You can see the forest plot clearly as shown in Fig. 12.27 and copy it from RevMan in word or any other file.
28. You can also copy the funnel plot as shown in Fig. 12.28.

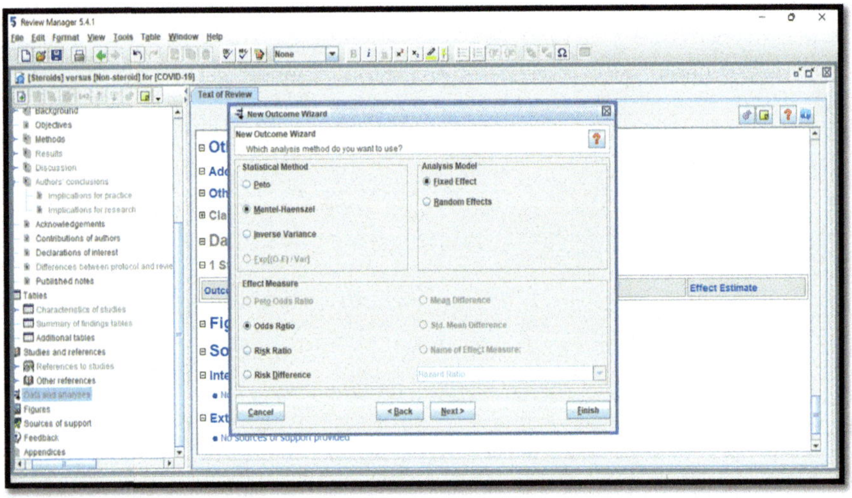

Fig. 12.21 Screenshot of selection of statistical method, analysis model, and effect measure

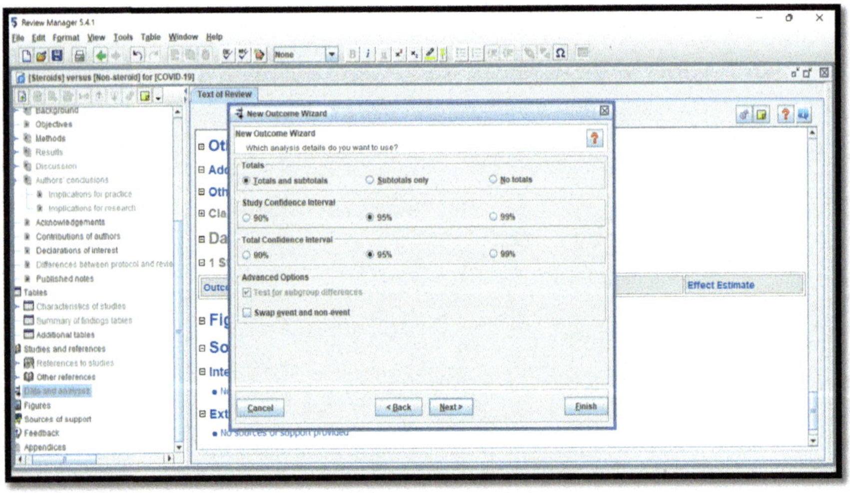

Fig. 12.22 Screenshot of selection of confidence interval

12.6.2 Comprehensive Meta-Analysis (CMA)

Comprehensive meta-analysis (CMA) is one of the popular packages which is user-friendly with excellent graphics. The data from each study were entered into a spreadsheet. It allows users to import data from Excel sheets directly. The program uses a spreadsheet for data entry but requires the user to identify specific columns to hold the study names and the effect size data. It accepts data in various formats and

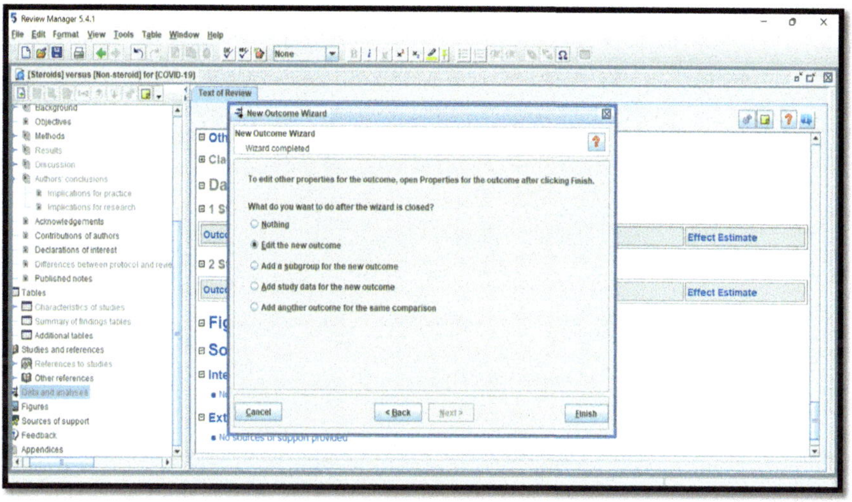

Fig. 12.23 Screenshot of completed new outcome wizard

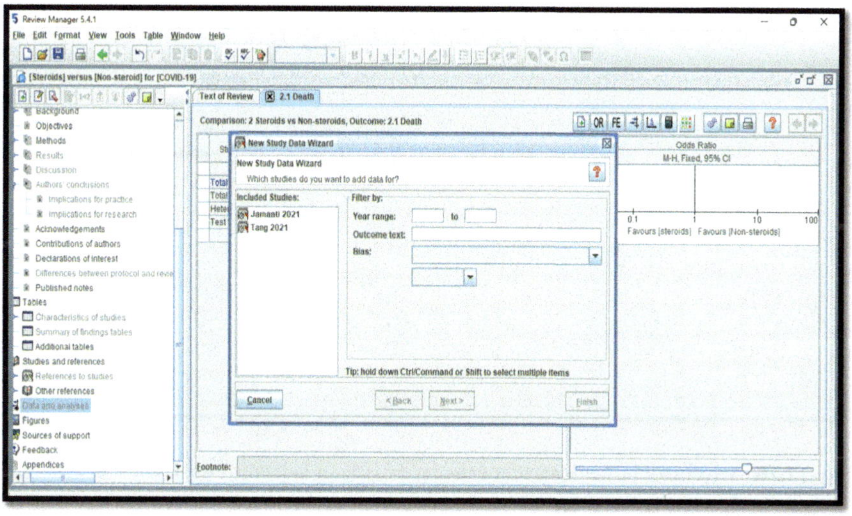

Fig. 12.24 Screenshot of New study data wizard

calculates overall estimates in terms of mean difference (MD), standard mean difference (SMD), odds ratio (OR), relative risk (RR), correlation, and hazard ratio. The quantitative analysis of publication bias was also done using Begg and Egger tests in addition to qualitative analysis of publication bias using a funnel plot. The user can also select any model as per the requirement of a particular study. The sensitivity analysis can also be done which includes an omit one analysis where meta-analysis is run repeatedly, with one study removed on each pass. However, it

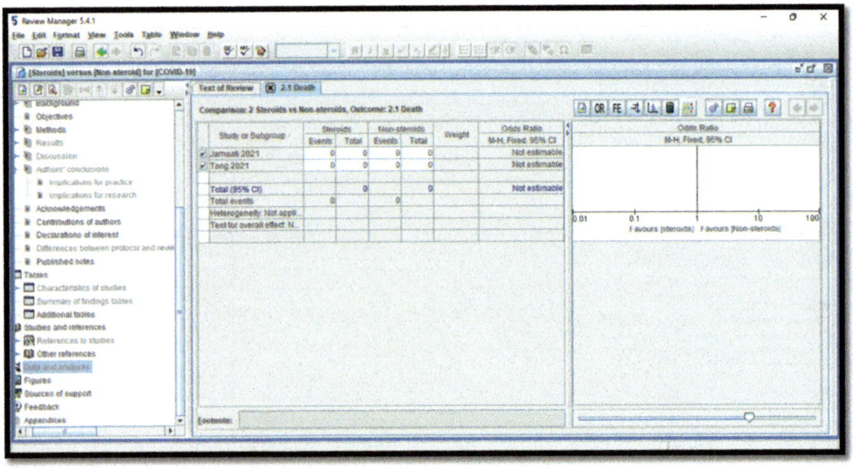

Fig. 12.25 Screenshot of tables

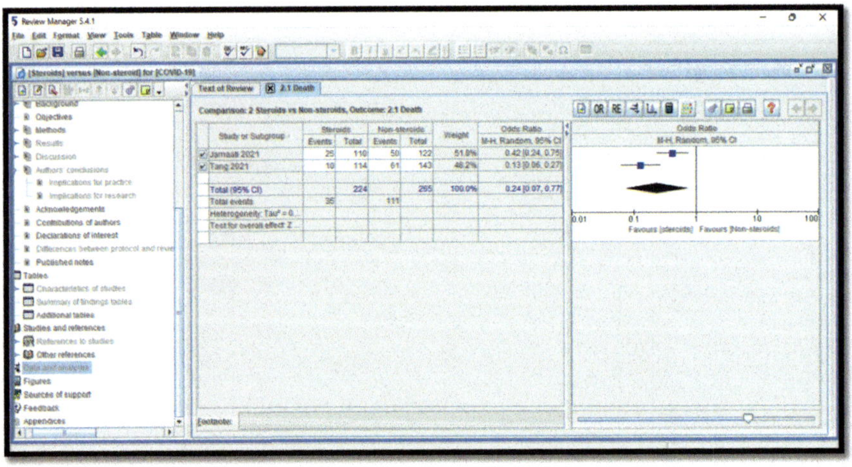

Fig. 12.26 Screenshot of RevMan after entering the data of individual studies

cannot perform multivariate analysis across time. The results of CMA can be exported to Microsoft Word. The CMA can be downloaded from www.meta-analysis.com as a demo for 10 days. The cost of CMA varies on the type of users (Faculty, Students, Corporate).

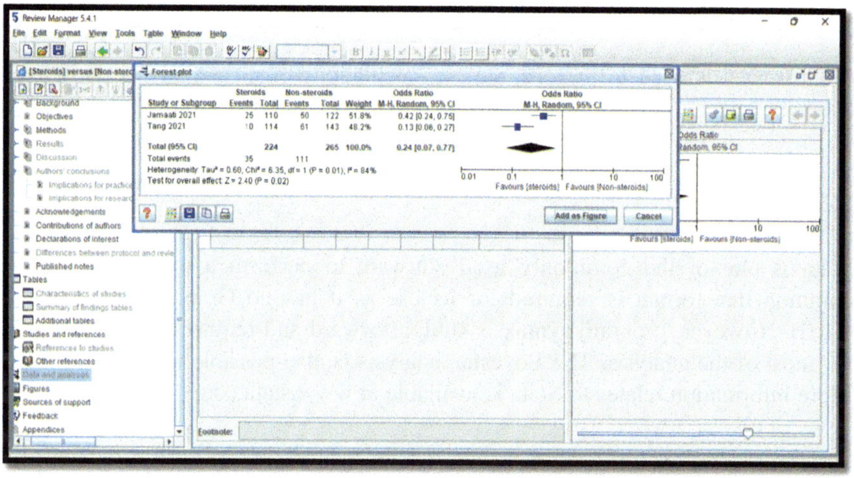

Fig. 12.27 Forest plot

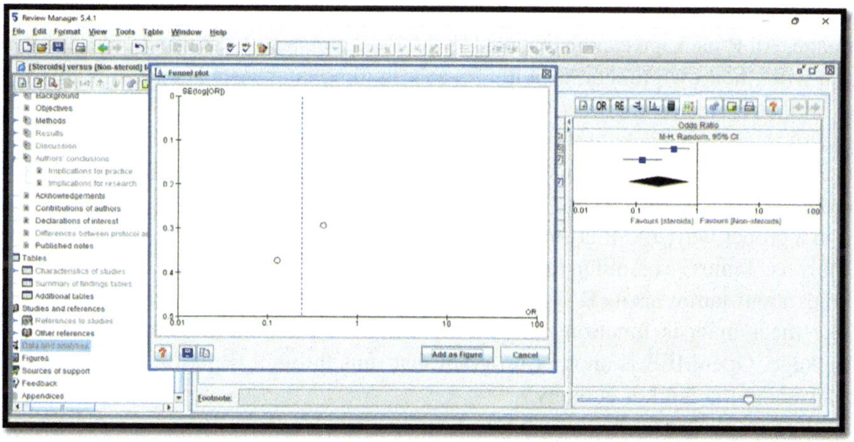

Fig. 12.28 Funnel plot

12.6.3 MetaWin 2.0 and PhyloMeta

MetaWin is specifically designed for the analysis of data produced from ecological studies. It is easy to use and contains special options like randomization and bootstrapping procedures. However, the complex meta-analyses modelling is not possible to perform using MetaWin. More information about MetaWin 2.0 is available at http://www.metawinsoft.com/. PhyloMeta is another package used for ecological and evolutionary meta-analyses.

12.6.4 MetaAnalyst

MetaAnalyst is used to perform simple calculations of meta-analysis. It is freely available and can be installed from http://tuftscaes.org/meta_analyst/.

12.6.5 Stata

Stata is one of the commonly used software to perform a meta-analysis. The common line format is required for its use as it has no Graphic User Interface (GUI). However, program syntax is straightforward and requires one line of code for most of the analysis. The Bayesian analysis is also possible to run using Stata. More information relates to Stata is available at www.stata.com.

12.6.6 R Packages

An r package is freely available for users which can perform many functions including meta-analysis. It is an open-source version of the S programming language. An R package can be installed in the local computing environment. The users can write their own functions in the R programming language. The meta-regression, as well as Bayesian analyses, can also be performed using the R package. Metafor is one of the best R packages used for meta-analysis. The codes are mentioned in the user guide which is very useful for analysis and creation of plots. However, R environment is a must to use R packages; therefore, users need time to learn to use it in a proper way. JASP is another package of R which is used to perform statistical analysis. Jamovi is similar to JASP that uses the Metafor package in R. The best thing about Jamovi is its R syntax mode, enabling the ability to see the used R code. The meta-analysis function of JASP is based on the aforementioned Metafor R package. OpenMEE is another program that runs through R, particularly for meta-analysis in the field of ecology.

R packages are freely available and can be downloaded from www.r-project.org.

12.6.7 SPSS

SPSS is the most widely used statistical package in the social sciences. These packages are also useful for calculations of random and fixed effect meta-analysis as well as meta-regression. However, command line interface should be used for these analyses instead of pull-down menus in SPSS.

12.6.8 Meta-Essentials (Excel Workbook)

Meta-Essentials is one of the simplest packages to perform a meta-analysis. You have to simply replace the example data which is entered in the Microsoft Excel workbook with your data and analysis will be conducted using simple steps. All analysis can be done within Excel.

12.6.9 MetaEasy (Excel Add-on)

MetaEasy is another Excel add-on that provides the ability to perform a simple meta-analysis.

12.6.10 MetaGenyo

MetaGenyo is an online interface used to perform meta-analyses on genetic association studies. The interface is very easy to use.

12.6.11 StatsDirect

StatsDirect is one of the most commonly used commercial software to perform a meta-analysis. It is exceptionally easy to use and calculate overall estimate measures in terms of odds ratio, relative risks, risk ratio, etc. for qualitative data, whereas MD and SMD are for quantitative data. No programming is required. It also calculates the overall prevalence. It also performs a quantitative analysis of publication bias. The trial version of StatsDirect is available for a limited duration for researchers.

12.7 Conclusion

In conclusion, a registry of systematic reviews may reduce publication bias, enhance transparency, and avoid duplication of effort.

Appropriate user-friendly software is required to perform calculations related to meta-analysis. Many commercially or freely available software perform these calculations; however, statistical packages require technical knowledge in programming. Therefore, many users struggle to use these packages in the correct way. In my personal view, stand-alone packages are more useful, particularly for non-technical users. The popularity of meta-analysis is increasing year by year; therefore, demand for software will continue to increase in the future.

References

Campbell Collaboration: A registry for systematic reviews of social interventions. https://www.campbellcollaboration.org/. Accessed 12 Feb 2022

International Platform of Registered Systematic Review and Meta-Analysis Protocols. https://inplasy.com/. Accessed 15 Feb 2022

Tawfik GM, Giang HTN, Ghozy S et al (2020) Protocol registration issues of systematic review and meta-analysis studies: a survey of global researchers. BMC Med Res Methodol 20:213

The Cochrane Collaboration (2020) RevMan 5 whenever its output is used: review manager (RevMan) [Computer program]. Version 5.4.1. Accessed 15 Feb 2022

The Cochrane Database of Systematic Reviews (CDSR). https://www.cochranelibrary.com/cdsr/about-cdsr. Accessed 12 Feb 2022

The International Prospective Register of Systematic Reviews (PROSPERO). https://www.crd.york.ac.uk/prospero/. Accessed 12 Feb 2022

Xu C, Cheng LL, Liu Y et al (2019) Protocol registration or development may benefit the design, conduct and reporting of dose-response meta-analysis: empirical evidence from a literature survey. BMC Med Res Methodol 19(1):78

Common Mistakes 13

Abstract

There are a handful of errors that arise on a pretty frequent basis among the thousands of meta-analyses that have been published over the last several decades. The most common mistakes which I personally observed are data entry errors/transposition errors, outliers, study ID appearing more than once in a forest plot, reporting at odds with forest plot, SEMs used instead of SDs, studies included multiple times, use of SMD and MD incorrectly, an adequate number of studies, selection of model based on results of heterogeneity, and so on. This chapter discusses the most common mistakes made by researchers in meta-analysis.

Keywords

Common mistakes · Meta-analysis

13.1 Introduction

A meta-analysis is an important analysis that synthesises data from individual studies to address a common research question. These analyses are helpful for healthcare professionals as well as for health policymakers to develop effective guidelines. Various user-friendly software are available which makes this analysis easy. However, it has been observed that various mistakes are made in already published meta-analyses (Eysenck 1994). Therefore, it is very important to know about common mistakes in meta-analysis so that researchers avoid these whenever planning to do this kind of analysis. In this chapter, I have highlighted several typical mistakes in systematic reviews and meta-analyses.

A. Kumar, *Meta-analysis in Clinical Research: Principles and Procedures*, https://doi.org/10.1007/978-981-99-2370-0_13

13.2 Common Mistakes in Meta-Analysis

Many systematic reviews have been published that do not adhere to the Preferred Reporting Items for Systematic Reviews and Meta-Analyses (PRISMA) standard regarding systematic review reporting quality. One of the most common blunders is considering Google Scholar as one of the databases available in the literature search area. It ought to be referred to as a 'search engine'. It is critical that the reviewers identify which is a database and which is a search engine. Other most common mistakes in the meta-analysis are mistakes in data entry errors/transposition errors, outliers, study ID appearing more than once in a forest plot, reporting at odds with forest plot, SEMs used instead of SDs, studies included multiple times, use of SMD and MD incorrectly, an adequate number of studies, random effects versus fixed effects, etc. (Shamsi et al. 2019, 2020; Easterbrook et al. 1991; Michael 2019). I have discussed each of these as mentioned below.

13.2.1 Data Entry Errors/Transposition Errors

The most common error is related to data entry. Therefore, it is always recommended to enter the data by two independent reviewers and finally do the cross-check.

13.2.2 Search Strategy

There are numerous errors in the search strategy. For example, the main objective of the study is to investigate the 'incidence'. However, I have found 'prevalence' in search items that should not be added.

13.2.3 Flow Diagrams

There are also common mistakes in the flow diagram representing the selection of studies. Sometimes, numbers are not matched correctly.

13.2.4 Quality Assessment

Another aspect to consider is importing the quality assessment of the evidence that has been incorporated (risk of bias within studies). The methodological quality evaluation is a critical notion; according to the PRISMA statement, the findings of assessing the methodological quality of each evidence should be included; nevertheless, in this form, the interpretation of results was presented without any attempt at the quality assessment.

13.2.5 Adequate Number of Studies

I have personally observed a few published meta-analyses that included two or three studies. An author should include at least 10 studies to draw a valid conclusion.

13.2.6 Forest Plot

Study ID appearing more than once in a forest plot, reporting at odds with the forest plot.

13.2.7 Selection of Effect Sizes

I have also observed some meta-analyses where the selection of a particular effect size is not clear. For example, researchers have used standard mean difference instead of mean difference for continuous data without any specific reasons, selection of odd ratio vs relative risk for categorical data, and so on.

13.2.8 Selection of Model

Most researchers provide an explanation for using a random effect model based on the results of heterogeneity, which is completely wrong. The author should select a particular model before running the meta-analysis based on variation among studies.

13.2.9 Interpretations of Results

Some authors interpret meta-analysis results incorrectly.

13.2.10 Other Issues

Apart from these above-mentioned mistakes, many are not possible to detect, like Have any papers been missed? Have the right results been copied from the papers? Have standard deviations been confused with standard errors?

13.3 Conclusion

In conclusion, the authors should be very careful whenever conducting a systematic review or meta-analysis on any topic of interest.

References

Easterbrook PJ, Berlin JA, Gopalan R et al (1991) Publication bias in clinical research. Lancet 337(8746):867–872

Eysenck HJ (1994) Meta-analysis and its problems. BMJ 309(6957):789–792

Michael B (2019) Common mistakes in meta-analysis and how to avoid them. Biostat, Englewood, NJ

Shamsi MB, Arab-Zozani M, Mirzaei M (2019) Methodological issue on reporting of systematic review of diagnostic accuracy of rapid ultrasound in shock. Bull Emerg Trauma 7(3):337–338

Shamsi M, Mirzaei M, Vaziri S et al (2020) Common mistakes in reporting systematic reviews and meta-analyses. Health Promot Perspect 10(2):97

Challenges

14

Abstract

Meta-analysis is always criticised for combining diverse types of studies ('mixing apples with oranges'). There are many challenges in the conduct of meta-analysis such as decisions regarding inclusion and exclusion criteria, availability of studies, quality of included studies, dealing with publication bias, and so on. In this chapter, I have mainly discussed common challenges in the field of systematic reviews and meta-analyses.

Keywords

Systematic review · Challenges · Meta-analysis

14.1 Introduction

The meta-analysis should be viewed as a framework for organising the steps involved in conducting a comprehensive assessment of previous research. The difficulties of evidence completeness and combinability, which must be considered in any review, are explicitly addressed. Was it a good idea to aggregate the individual trials that made up the meta-analysis? What is the result's resiliency to changes in assumptions? Is the conclusion clinically sound? Finally, how has the research aided in the process of making reasonable judgements concerning patient care?

The 'apples and oranges' criticism is always there which refers to the improper mixing of research within a single comparison with the goal of estimating an average impact size. The different formulations of the same intervention can behave differently in different studies and meta-analysis of these studies will not be useful as an aggregation of results from quite different formulations may be suspect. The decision regarding the suitability of studies to perform meta-analysis is not always clear as it can be influenced to some extent by the issue being addressed. The decision is

A. Kumar, *Meta-analysis in Clinical Research: Principles and Procedures*,
https://doi.org/10.1007/978-981-99-2370-0_14

sometimes referred to as 'lumping' (when studies are combined in the same analysis) or 'splitting' (when they are not) (Metelli and Chaimani 2020). The current issues in the field of the meta-analysis are briefly discussed in this chapter.

14.2 Challenges

Perhaps the most difficult aspect of the project is deciding which papers to include. I am referring to the choice between studying a very particular, limited question (for example, the same medicine, sickness, and setting for research following the same protocol) or a broader general one (e.g. a broad class of treatments for a range of conditions in a variety of settings). The more comprehensive a meta-analysis is, the more difficult it is to interpret the pooled information in terms of future policy. For example, researchers have concluded a statistical association of drug with a particular outcome based on results of a meta-analysis and suppose included individual studies differ significantly in design, it would be a mistake to believe that the results of this meta-analysis are valid in all cases. As a result, the larger the group, the better. Therefore, the results of any meta-analysis should be approached with caution.

The quality of data clearly influences the trustworthiness of a meta-analysis. In my opinion, one should look for adherence to studies with protocol and any deviations from protocol should be carefully considered. In clinical trials, one should look for randomization, blinding, analysis, etc. to ensure the quality of data. To ensure the quality of data of individual studies considered for meta-analysis is the most difficult task and failure to identify all qualifying studies, whether owing to publication bias or missing published studies, poses a substantial danger of data distortion. The overall quality of meta-analysis is increased if researchers have assessed the quality of data. Therefore, I always suggest my students to assess the quality of data of included studies for meta-analysis.

When it comes to the methodologies for calculating an overall assessment of the amount of effect, there is more debate. The selection of the right model is also very important. The 'fixed effect' model is the most prevalent as it calculates the weighted mean of the individual studies which is inversely proportional to sampling variances. The overall estimate measures are always represented with an overall confidence interval which is invariably narrower than the interval for any individual study. In most of the published meta-analyses, the fixed effect model is used less as compared to the random effect model. The possible reason might be an assumption of a fixed effect model, i.e. no heterogeneity among studies and all individual studies are estimating the same true effect and only differ due to sampling variance. However, most of the time, the individual studies differ in many ways, therefore, this assumption is improbable in most meta-analyses. The most of tests used for the calculation of heterogeneity lack power and it is always recommended to choose a model based on personal judgement. If tests show non-significant heterogeneity, one cannot say that the fixed effect technique is valid; all one can say is that there is no large heterogeneity. The clinical heterogeneity (i.e. changes in study design) is common, and establishing what the fixed effect technique is truly assessing can be problematic. The confidence interval for 'fixed effects' only reflects the uncertainty due to sampling variance.

The 'random effects' model considers statistical heterogeneity in the overall estimate of an average effect. However, I believe this method has both conceptual and practical flaws. This method assumes that studies are a random 'sample' from a hypothetical 'population'; therefore, estimation of between-study variance in true effects will usually be imprecise. Further, if heterogeneity exists, it assigns more weightage to smaller studies as compared to the fixed effect method, despite the fact that smaller studies are often of lower quality. Is the random effects technique (i.e. are the findings the same?) a good illustrative supplement to the fixed effect method? However, until there are a lot of variabilities, I prefer the latter.

If statistical heterogeneity is observed, it appears that investigating the causes of the heterogeneity is more essential than modelling it using random effects. Thus, one must analyse which aspects of the studies or the individuals within the studies might explain why heterogeneity occurs, either informally or by statistical analysis, and this is likely to be most fruitful if raw individual data is available.

One critique of meta-analyses is that sometimes researchers tried so hard to combine data from different studies into a single estimate. In my opinion, researchers should be self-critical regarding the limitations of meta-analysis techniques. Both approaches (random or fixed) can neither be properly tested. The competent and thorough meta-analyses are significant efforts that have had a significant impact on health policy.

If researchers would like to check the effect of a particular treatment on a particular outcome, then the important question of how to choose which trials to include in the meta-analysis arises? In an ideal world, all relevant trials, whether published or not, would be included.

Positive trials are heavily favoured in the medical literature. The reasons might be due to the editors. The editors of most of the journals are more interested to accept positive findings instead of negative findings (Norbert 1995; Vo and Vansteelandt 2021; Cashin et al. 2019).

14.3 Criticism in Meta-Analysis

Meta-analysis has been criticised for combining diverse types of studies ("apples and oranges"). Nonetheless, by examining the heterogeneity of the research and performing subgroup analysis, this difficulty can be solved. A meta-analysis should be avoided if studies are too varied to be comparable, as the meta-analysis result may be meaningless and any genuine effect may be disguised. Another restriction of meta-analysis is the principle of "garbage in, garbage out," which states that if a meta-analysis includes low-quality research with bias, the results will be prejudiced and erroneous. As a result, the meta-results analysis's are reliant on the quality of the main study. As a result, every meta-analysis should include trials that have been carefully chosen based on tight inclusion criteria.

14.4 Limitations

It is important to realise that meta-analysis has some substantial drawbacks. The most significant of them, in my opinion, is the complete reliance on the quality of the papers included. No amount of statistical manipulation can compensate for flaws in studies that were poorly done, a principle known as garbage in–garbage out.

Another key issue is the heterogeneity of the investigations. The number and kind of patients involved, the study's location, the procedures and/or equipment employed, the investigators' experience, the presence of supplementary drugs, and other factors can all contribute to heterogeneity. Individual studies may, however, yield different outcomes. Obviously, the more consistent the conclusions of individual studies are, the more confident one can be in the integration results (Higgins et al. 2019).

The impact of publication bias, or the tendency for research with negative results to be submitted and accepted for publication less frequently than studies with good results, is particularly vulnerable in meta-analysis. Hopefully, early registration of clinical investigations will aid in overcoming this issue. Finally, meta-analysis, like other research endeavours, is vulnerable to the investigator's and/or sponsor's possible bias.

In addition to the aforementioned constraints, meta-analysis, like any research, is prone to methodological problems. A comprehensive literature search is required for accurate meta-analytic methods, as are precise criteria for the removal of studies. Meta-analysis is generally rated based on the quality of the data included, the quality of data extraction, the statistical methods used to integrate data, and the degree to which the study's shortcomings are acknowledged and addressed.

Despite these well-defined conditions, there is evidence that meta-analyses are frequently carried out incorrectly.

Given the foregoing, it is not surprising that many people have questioned the utility of meta-analysis. They doubt that various studies involving different patients from different locations who were treated for different lengths of time according to different methods could provide any relevant results.

In terms of meta-analysis, I have conflicting feelings. All of the appealing features of these methodical, quantitative reviews are obvious to me. No one can deny the importance of a publication that identifies the best data on a topic, applies valid statistical analysis to the extracted data, and draws a clear conclusion from the whole that is not supported by the individual research. These papers can help you stay current with the literature and lay the groundwork for evidence-based medical practice.

I am concerned that such studies encourage me to be intellectually lazy and avoid grappling with the underlying material on my own, but this is not the study's fault. However, I feel that this technique can only be applied in a limited number of situations where fresh, valid, and meaningful information can be produced. When there is a sufficient number of reasonably well-done, usually homogeneous randomised trials but not enough power to produce a statistically significant result, the ideal scenario exists. When there are only a few small trials available or when there are a large number of trials with the same results, meta-analysis is less useful. If the data is not dissimilar, why bother with a meta-analysis?

When there are several studies with conflicting results, a meta-analysis may be useful. In this case, however, the variability of limits and a large randomised clinical study would be a preferable way to tackle the problem.

14.5 Conclusion

Finally, despite its flaws, meta-analysis is popular among readers, highly welcomed by reviewers, and is increasingly employed to create evidence-based practices and guidelines. Although it defies logic to believe that quantitative integration of a lot of disparate investigations may yield new, reliable information, the numerous meta-analyses published speak for themselves. When used properly, I believe they can make a significant contribution.

References

Cashin AG, Lee H, Lamb SE et al (2019) An overview of systematic reviews found suboptimal reporting and methodological limitations of mediation studies investigating causal mechanisms. J Clin Epidemiol 111:60–68

Higgins JP, Thomas J, Chandler J et al (2019) Cochrane handbook for systematic reviews of interventions. Wiley, New York, NY

Metelli S, Chaimani A (2020) Challenges in meta-analyses with observational studies. Evid Based Ment Health 23(2):83–87

Norbert V (1995) The challenge of meta-analysis: discussion. Indications and contra-indications for meta-analysis. J Clin Epidemiol 48:5–8

Vo TT, Vansteelandt S (2021) Challenges in systematic reviews and meta-analyses of mediation analyses. arXiv preprint arXiv:2103.12227

Future Perspectives 15

Abstract

Recent advancements in the field of artificial intelligence and machine learning could play an important role in the field of meta-analysis by reducing the manual work of systematic review of the literature. This chapter provides a brief introduction to future perspectives in the field of meta-analysis.

Keywords

Artificial intelligence · Machine learning · Meta-analysis

15.1 Introduction

Artificial intelligence (AI) is defined as "the science and engineering of making intelligent machines, especially intelligent computer programs". Machine learning (ML) is a subset of AI in which the algorithm allows computer programs to improve without additional programming. These techniques are playing a significant role in various fields, including the field of health. In this chapter, I have briefly discussed the role of such types of techniques in the future in the field of meta-analysis.

15.2 Future Perspectives

In recent years, there has been a lot of interest in using machine learning and text mining to automate the study selection process, which has resulted in the development of a variety of tools and methodologies for reviewers to examine. The use of automated techniques has the potential to greatly minimise the burden associated with study selection.

A. Kumar, *Meta-analysis in Clinical Research: Principles and Procedures*, https://doi.org/10.1007/978-981-99-2370-0_15

Recently, Michelson et al. (2020) have demonstrated a novel approach to meta-analysis using artificial intelligence (AI) to generate meaningful clinical insights in a much shorter time period than traditional meta-analysis (Michelson et al. 2020). In the future, machine learning models can also be built where sufficient data are available and one can identify the reports of randomised trials based on titles and abstracts. It will reduce the errors and provide studies in a quick manner with high accuracy. The models could flag excluded studies as per the inclusion and exclusion criteria of the study. For example, normally case reports, already published reviews, non-randomised studies are excluded so these studies can be flagged up in the model. Overall, machine learning models will learn how to apply particular eligibility criteria of the particular study. Some software tools that help the authors during the screening process are Abstrackr, Colandr, EPPI-Reviewer, Rayyan, RobotAnalyst, Swift-review, and so on. Abstrackr is a free web-based screening application that uses machine learning approaches to prioritise record screening (Wallace et al. 2012). Covidence is a web-based software platform for a systematic review that incorporates collaborative title and abstract screening, full-text review, risk-of-bias assessment, and data extraction. The authors of Cochrane Reviews get free access to this system, which ordinarily needs a paid subscription. There is also a free trial for non-Cochrane review authors (Covidence 2022).

Recently, Schoot et al. (2021) have also designed an open-source machine learning framework to accelerate the step of screening titles and abstracts (Schoot et al. 2021). Xiong et al. (2018) also created a novel machine learning approach for identifying papers that are eligible for meta-analysis (Xiong et al. 2018). Panahi et al. (2021) have used machine learning approaches to conduct meta-analysis (Panahi et al. 2021).

15.3 Conclusion

In the future, I believe that machine learning and artificial intelligence will also play a significant role in the field of meta-analysis as well.

References

Covidence. https://www.covidence.org/. Accessed 10 Jan 2022

Michelson M, Chow T, Martin NA et al (2020) Artificial intelligence for rapid meta-analysis: case study on ocular toxicity of hydroxychloroquine. J Med Internet Res 22(8):e20007

Panahi R, Ebrahimie E, Niazi A et al (2021) Integration of meta-analysis and supervised machine learning for pattern recognition in breast cancer using epigenetic data. Inform Med Unlocked 24: 100629

Schoot RV, Bruin JD, Schram R et al (2021) An open source machine learning framework for efficient and transparent systematic reviews. Nat Mach Intell 3:125–133

Wallace CB, Small K, Brodley CE et al (2012) Deploying an interactive machine learning system in an evidence-based practice center: abstrackr. In: Proceedings of the ACM International Health Informatics Symposium (IHI), pp 819–824

Xiong Z, Liu T, Tse G et al (2018) A machine learning aided systematic review and meta-analysis of the relative risk of atrial fibrillation in patients with diabetes mellitus. Front Physiol 9:835

The manufacturer's authorised representative in the EU is Springer Nature Customer Service Centre GmbH, Europaplatz 3, 69115 Heidelberg, Germany. If you have any concerns regarding our products, please contact ProductSafety@springernature.com

Printed and bound by CPI Group (UK) Ltd, Croydon, CR0 4YY
01/10/2026
02225877-0005